THE HUDSON RIVER HIGHLANDS

Artist unknown, *Sailboats and Sidewheeler "Star,"* undated mid-nineteenth century oil on canvas, showing the passage through the Highlands of the Hudson known as the "Wey-Gat" or "Wind-Gate." Collection of the Hudson River Museum, Yonkers, N.Y. Anonymous donor.

THE HUDSON RIVER HIGHLANDS

FRANCES F. DUNWELL

Foreword by Robert F. Kennedy, Jr.

Columbia University Press ~ NEW YORK

The views expressed herein do not necessarily reflect the beliefs or opinions of the sponsors or the donors, who assume no liability for the contents or the use of the information herein.

COLUMBIA UNIVERSITY PRESS

New York Chichester, West Sussex

Library of Congress Cataloging-in-Publication Data

Dunwell, Frances F.

The Hudson River highlands / Frances F. Dunwell ;

foreword by Robert F. Kennedy, Jr.

p. cm.

Includes bibliographical references and index.

ISBN 0-231-07042-X (cloth edition; alk. paper)

0-231-07043-8 (paperback edition)

1. Hudson Highlands (N.Y.) I. Title.

F127.H8D86 1991

974.7'3—dc20

90-26674

CIP

Book design by Jennifer Dossin

Casebound editions of Columbia University Press books are printed on permanent and durable acid-free paper

Printed in the United States of America

c 10 9 8 7 6 5 4 3 2 1

p 10 9 8 7 6

TO MY FAMILY AND FRIENDS

and, in particular, to Franny Reese, without whom, for so many reasons,
this book could not have been written

and to the Hudson,
river of mystery and mountains

Contents

Foreword

WHEN I WAS a child my father took me and eight or nine of my brothers and sisters white water rafting near North Forks in the Adirondacks; the water was so clean it tempted me to drink it, but the guides said it was better not to. I thought about that often in the ensuing years, and in 1984, I had the opportunity to do something about it. That year I began working as an attorney for the Hudson River Fishermen's Association in Garrison. We worked out of a small farm house below Osborn's castle in the river stretch between Bear Mountain and Storm King. It was then also that I fell in love with the Hudson Highlands rising in fairy tale splendor from the river's banks. The grand homes of the nineteenth-century railroad robber barons seem to nestle in every gully between these Sugarloaf peaks. The Fishermen operate a patrol boat, the *Riverkeeper,* which plies the river searching for polluters.

The Hudson is home to one of the most vigilant, sophisticated and aggressive environmental communities to protect any resource in the world. Since the early 1960s an uncompromising collection of environmental associations: the Natural Resources Defense Council, the Hudson River Fishermen's Association, Clearwater and Scenic Hudson, to name a few, have fought to maintain the river's biological and aesthetic integrity.

Today there is a citizen group at every bend and on both banks ready to do battle with any potential polluter or developer. Many of these battles have ended in victory. During the past decade and a half, Hudson River environmentalists have succeeded in stopping the construction of two major Hudson River highways, two nuclear power plants, and the proposed pumped storage facility on Storm King Mountain. We have forced one utility to create a $17 million endowment for the river and to spend another $20 million to research and build special equipment to avoid fish kills at its power plants. In the past five years, the Hudson River Fishermen's Association has brought more than 40 polluters to justice. Their willingness to hire lawyers to fight has contributed substantially to our country's environmental case law. Successful cases against Exxon for its secret water thefts, the Westway project, General Electric for its PCB discharge, Anaconda Wire and Copper for its toxic discharges, and Con Ed for fish kills at its Indian Point Power Plant, have created a black letter jurisprudence that forms the foundation of American environmental law. The Hudson River Storm King case is the first case in many environmental law textbooks. That decision gave American environmentalists the standing to sue polluters who injured their aesthetic values. Prior to Storm King, environmentalists could bring actions only where they could prove direct economic damage.

The fight to preserve the river's viewshed and its astounding biological productivity continues. Once the butt of *Tonight Show* quips, the Hudson surprisingly is, gallon per gallon, among the planet's most productive water bodies. The Hudson is the last estuary on the East Coast of North America and perhaps in the entire North Atlantic drainage that still retains strong spawning stocks of all its historical fish species. I have seen the evidence of this productivity first hand.

Although the Hudson itself is too turbid for good visibility, I've often gone scuba diving in its tributaries. My favorite is the Croton River where I've snorkeled or scuba dived from the rail yard at its mouth, upstream to Quaker Rock below the first dam. The pool there is about 12 feet deep during high water in the spring. If I go to the bottom and hug a big rock to stabilize myself in the current, I can sit and watch the pool fill up with fish. Thousands and thousands of alewives or "sawbellies" come up the Croton on their spawning run from the Atlantic in the early spring. You can see great striped bass that swim up to feed on the herring schools or to spawn. Just above the railyard the Croton River bottom is carpeted with barnacles. Blue crabs are common during some seasons and you see glass shrimp and other marine invertebrates and crustaceans almost anytime during the spring or summer. Atlantic eels 5 feet long hide between the rocks and crevices, having migrated from their spawning grounds in the North Atlantic. They feed in the Hudson River for a couple of years and then run back to the Sargasso sea to spawn. Ravenous schools of snapper bluefish, which I used to catch on Cape Cod as a boy, snatch mayflies in their vicious little teeth, competing with the Croton's abundant trout populations—rainbow, brown, and brook. The freshwater species—largemouth bass, smallmouth bass, white perch, yellow perch, bluegills, beautiful pumpkin seed sunfish that are as colorful as any of the tropical fish that adorn pet store fish tanks—congregate in large numbers in the aerated waters beneath the spillway.

Because of its hydrological connection to the Gulf Stream, tropical fish are common in the estuary. In the Croton, one sees odd creatures like seahorses and a fish called a moongazer, which emits an electric shock when you touch it. Jack Crevalle and Atlantic needle fish, whose home waters are in Florida and the Carribean, are abundant in the Hudson. Giant carp, weighing 30 to 50 pounds and having the appearance of giant mustachioed goldfish, darken the bottom as they swim overhead in pairs or trios. We also have proper goldfish in the Hudson, though they are less abundant than in former times. They grow extraordinarily large—up to four pounds—and at least one man, Everett Nack of Claverack, New York, supplements his income by capturing them for collectors.

The underwater observer is struck by the Hudson's diversity and by its abundance; the impression is of being thrown into a stocked pet store aquarium. Hudson fisherman Tom Lake wrote me in a recent letter that he had caught a permit in the Hudson, the river's 200th confirmed species. Two years ago last March, I was out with a commercial fisherman under the Tappan Zee Bridge pulling up ten thousand pounds of fish on every lift. The nets were filled with each tide every six hours, day and night, week after week, month after month, all springtime. The State Department of Environmental Conservation asked the commercial fishermen to drive their boats slowly in the shallows because at high speeds they were hitting so many fish. This spring the fishermen pulled their nets early because the bountiful striped bass fouled their nets making it unprofitable to fish for shad. Striped bass from the Hudson cannot be sold, because they are contaminated with PCBs.

Because of its relative diversity, health, and abundance, the Hudson is the Noah's Ark of the East Coast and perhaps for the entire North Atlantic. As the great estuaries of the North Atlantic—the Chesapeake, the St. Lawrence, the Sea of Isov, the Rhine, Long Island Sound, Narragansett Bay, and the Carolina estuaries—decline, the Hudson grows in importance. It is the gene pool—the last safe harbor for species that face extinction elsewhere.

The Hudson's defenders have won many battles to save the river, but ironically, our success on the Hudson has increased the river's allure to developers and industry. The most challenging fights are yet to come. The big challenge to the Hudson's viability today comes not from a single giant project but the death of a thousand cuts, the decline of the river as condominiums occupy every available piece of shoreline.

Sometimes in a court, or across a negotiating table from a polluter or developer, I am asked of the Hudson or perhaps of one of its denizens, "How much is it worth?" In one context we have the tools to measure the value of, say, the Hudson River striped bass. We can estimate their numbers and multiply by twice the cost of a striped bass filet at the South Street fish market. Or we can derive that value through the dollars paid the state for fishing licenses or the total revenues of local bait shops and marinas. But there are other costs we cannot measure—the cultural and spiritual costs of losing a species. We have a right, even an obligation, to make economic choices for our children—but there are other choices we have no right to make and there are costs we have no right to impose. I brought the question to Thomas Berry, a Catholic priest who lives on the Hudson and contemplates our environment, "Why do we need to spend all these resources saving, for example, the snail darter?" The snail darter is a small fish whose presence in a Tennessee river delayed a million dollar dam project. Unlike the Hudson River striped bass, this fish is insignificant economically. From an ecological standpoint it is nearly insignificant; a dozen other members of the darter family could easily fill its ecological niche. He replied that if we lose a single species, we lose part of our ability to sense the divine, to know who God is. That is an old theological theme. St. Paul said we know the invisible God by looking at visible nature. St. Thomas Aquinas said that by looking at a single flower we can begin to think about God but that in order to comprehend something of the Divine majesty we must look at the entire fabric of nature with all its intricacies, its vast interwoven complexity.

Today as we pull the threads from that fabric and as the patchwork tatters, we lose part of our ability to "sense the Divine." This is a cost we cannot measure. The Hudson is a critical piece of that fabric. Under tremendous development and industrial pressure, the river has continued to thrive, partly as the result of its own miraculous resilience, and partly because of the devotion and sacrifice its beauty has inspired in the populations that live along its banks.

Fran Dunwell has given us a wonderful story explaining this symbolic power that the region has acquired over its people and over our nation. She shows us how we as Americans have responded to our environment in a place—the Hudson Highlands—where Divine presence has long been sensed. She traces the evolution of the conservation movement from the early roots of our culture to the development of an ethic which has spawned an environmental jurisprudence now protecting our resources from sea to sea. The Hudson Highlands have changed the way our nation views its land and natural resources and given us the tools to protect our environment as well as our spiritual and cultural heritage. We owe Fran our gratitude for capturing all this in her beautifully written book.

ROBERT F. KENNEDY, JR.
Mt. Kisco, New York
August 1991

Preface

I FIRST BEGAN thinking about writing a book on the Hudson River Highlands in 1978. The occasion was Sun Day, and I was one of about 200 people gathered for a celebration breakfast on the sandy beach of Little Stony Point, part of Hudson Highlands State Park in Cold Spring, New York. It was a brisk spring morning when the cold was wet and penetrating. We made pancakes and bacon on an open grill, our high spirits and cameraderie a warm contrast to the chill wind. The event was a fund-raiser, one of many, and took place in the fifteenth year of what would eventually be a seventeen-year crusade to halt the construction of a power plant on Storm King Mountain. Across the river, Storm King towered over us, cold and white in May's morning light. The Hudson is narrow here, and Storm King, although a relatively small mountain of 1300 feet, seemed monumental.

At the time, I was the director of a newly created environmental group called the Center for the Hudson River Valley and a newcomer among old-timers. This was my initiation and my first experience of the mountain. True, I had seen Storm King many times since childhood—but this was the first time it had touched my soul. From where we stood at the water's edge, it dwarfed us, awesome and magnificent—a raw element of the earth, created of fire and stone, swelling from the water toward the sky. It loomed over us, massive and

Storm King Mountain. Photograph copyright © by Raymond O'Brien.

mysterious. It was the moment when I understood why the people gathered there could have fought for so long to protect it.

In the Highlands, the powerful forces of nature are ever-present. It is a place that affects you emotionally, opens the senses, and connects you with the ground on which you stand. I spoke to many people that day, and I learned that the emotional power of the Highlands has been the source of a rich cultural history. The response to nature here has always been strong, and for centuries, Storm King and the other burly mountains which form these Highlands—Dunderberg, Crow's Nest, Breakneck, Anthony's Nose, Bear Mountain, and Sugar Loaf—have awakened people's passion and repeatedly called forth dramatic, sometimes heroic, response. The way men and women have related to the Highlands has changed from generation to generation, but each experience has added depth and meaning to the next. The fight to save Storm King, which ultimately drew the support of 20,000 people from 48 states and three continents, was the culmination of a centuries-old tradition.

The Hudson Highlands, a stretch of river only fifteen miles long, is the passage where the Hudson cuts through the Appalachian chain of mountains about fifty miles north of New York City, between Peekskill Bay and Newburgh Bay. To the north and south, the Hudson is straight and broad, but from the time it enters the Storm King gorge until it emerges through the "southern gate" by the Dunderberg, it is fast, narrow, winding, and deep. From either shore, the hills rise up more than 1000 feet, creating a landscape which is both beautiful and mysterious, a place of many moods, inspiring yet terrifying. So strong is the image of the Highlands that the area has come to symbolize the entire Hudson River, appearing, for example, as the background in a nineteenth-century painting of the Battle of Saratoga, which took place over two hundred miles to the north.

Feared by Dutch sailors, embraced by romantics, and captured on canvass by artists of the Hudson River School, the Highlands have cast a spell. Revolutionary generals stopped to admire the scenery as cannons exploded about them. The romance of the region inspired the folk tales of Washington Irving, the designs of architect A.J. Downing, and the castles of millionaires. Nineteenth-century seekers found spiritual renewal in this untamed wilderness. Among the railroad barons and other captains of industry, the desire to protect this landscape sparked extraordinary acts of philanthropy and the creation of vast parks. These mountains have witnessed monumental brilliance in the engineering of delicate bridges and scenic roads carved out of sheer cliffs at fantastic expense. The Highlands of the Hudson have played a special role in the social, cultural and political movements through which our nation has grown, shaping them and in turn being shaped by them. Their history is a story of the power of place and of human relationships with the land. The Highlands offer a place to awaken one's spirit and a connection with our roots as a nation. It is this which moved so many people through the centuries.

My experience at Little Stony Point was the beginning of a personal journey culminating in this book. I felt that the story of the Highlands was important to share with others, because it tells us something about ourselves and shows that another way of looking at things is possible. The Highlands remind us that nature and beauty are essential parts of our lives to be protected and preserved, and they demonstrate that we can do this if we only choose to try.

After the celebration breakfast, I had many occasions to revisit the Highlands, including New Year's Day of 1981, when a small hardy group climbed Storm King to celebrate the successful conclusion of the battle to save it. That year, I went to work for Scenic Hudson, the group organized in 1963 to preserve the Hudson River and Storm King. One of my assignments was to study the cottages, estates, barns, and bridges in the Highlands region with an eye toward putting qualifying structures on the National Register of Historic Places. Through this project and the team of people I worked with, I realized that history has left an uncommon imprint on the Highlands. Looking at the buildings, the patterns of development, the public works projects, and the parks was like reading about the people who lived there. They speak of an unusual harmony of humankind with nature. Much of the Highlands' landscape has been preserved in its wild state; and where there has been human activity, the built environment shows an unusual respect for nature, the result of conscious efforts to be a part of rather than to dominate the land. A journey through the mountains takes you down pre-Revolutionary dirt roads, past gabled toll houses and picturesque country homes, by old stone walls and white picket fences and into villages whose highest point is still the church steeple. Dense forest alternates with open vistas as you descend from the mountains to the bluffs, to the river. Seen from a boat on the river, the effect is even more dramatic. Even the vast military academy at West Point has been designed to complement the forms and textures of the land.

Despite this appearance of timelessness, progress did not evade the Highlands. For at least 150 years, the Highlands have been the scene of commerce and development. At times these uses coexisted peacefully with the enjoyment of the region's natural and cultural heritage. In the nineteenth century, mining, quarrying, clearcutting, and iron manufacturing advanced side by side with the blossoming of an artistic and spiritual movement which sought fulfillment in the unspoiled mountain landscape. More recently, the area was the scene of dramatic battles between competing interests, and today, the Highlands' landscape is important both for what is there and what is not.

In the Highlands you will not find a quaint, rural village plucked out of the New England countryside and deposited into the New York metropolitan area. There are spots of strip development and seedy deterioration mixed in with the region's rustic beauty. But throughout, the presence of nature is felt.

This book tells a story of nature and Americans beginning with the earliest colonial exploration. The traces of our experience with this landscape can be found in place names penned on early maps, in navigation charts, in military journals, personal diaries, works of art and literature, legislation, and in legal documents. They tell us the facts as well as the perceptions, for the way we relate to a place has a great deal to do with who we are, what we are doing there and what we expect. These records show an evolution of feelings toward the Highlands and chart the influence of the land on the course of national events. Above all they show that certain places have a special meaning in our lives; and they prove that citizens can protect what they passionately care about.

The modern environmental movement was born from the Storm King battle, and with it the activism of people who understood that the stewardship of the earth depends on the commitment and work of all of us as individuals. For many, this had been their first time speaking out, breaking ranks, putting themselves on the line to fight for something they believed in. The Storm King battle was a creative force, molding human courage. The spirit of the Highlands endures because people intervened to protect nature, beauty and a national heritage. The story of the Highlands is a tale of how that came to pass.

FRANCES F. DUNWELL
New Paltz
August 1991

Acknowledgments

THIS PROJECT HAS been co-sponsored by the Alice and Hamilton Fish Library and Hudson River Heritage. Scenic Hudson played a significant role in the early development of this book, supporting both its spiritual and temporal needs.

The author, the publisher, and the sponsors would like to take this opportunity to thank the following people and organizations which provided critical financial support. They are: the Gannett Foundation, the Hudson River Foundation, the J. M. Kaplan Fund, the Richard King Mellon Foundation through the Yale School of Forestry and Environmental Studies, the Newington-Cropsey Foundation, the Ogden Foundation, the Perkins Foundation, Alex Reese, Franny and Willis Reese, Constantine and Anne Sidamon-Eristoff, and Nancy Stover in memory of Robert Stover.

We also wish to express our gratitude to the New York State Council on the Arts. Through grants to Scenic Hudson in 1981 and 1982, NYSCA launched this project providing funding for pre-production research as well as photography by Robert Beckhard, whose lovely contemporary images of the Highlands illustrate the book.

In recognition of their early and most helpful support and to assist them in their important on-going programs to protect the Hudson River estuary and its shores, royalties earned from sales of this book will be donated to Scenic Hudson and the Hudson River Foundation.

IN ADDITION, the author would like to thank the following people for what can only be described as extraordinary effort and helpfulness without which this book would not have been possible:

Chuck Adams, Wint Aldrich, Deborah Begley, Dena Belzer, Betsy Blair, Francie Brody and her firm, Brody and Weiser, Lynda Cassanos, Bob Diamond, Julia and Steve Dunwell, Sidney Forman, Paul Keller, Cara Lee, Edna MacMahon, Lang Marsh, Carl Mondello, Leo Opdycke, Pamela Read, Franny Reese, Klara Sauer, Tamara Watson, Ellen Weiner and Bob Weinstein. This book is as much their product as mine.

Books such as this cannot be written without reference to historic collections of books, periodicals and documents. The following institutions and Libraries have outstanding collections which were of invaluable use:

USMA West Point, Franklin D. Roosevelt Library, Mid-Hudson Library System, New-York Historical Society, New York Public Library, New York State Library, NYS Office of Parks, Recreation and Historic Preservation, Putnam County Historical Society, Smithsonian Institution, SUNY New Paltz, and Yale University.

It has been nothing short of a pleasure to work with the staff and editors at Columbia University Press. They have made difficult tasks seem easy, they have been responsive and enthusiastic, and they have been unflappable in dealing with the unique complexities of this project. Special thanks to: Jennifer Crewe, Jennifer Dossin, Leslie Bialler, Rob Lindsay, Kathleen Szawiola, Kris Fresonke.

Early drafts of this book were written, typed and photocopied when I was a staff member at Scenic Hudson. For their tireless assistance I would like to thank: Jean Barker, Katherine Berkery, Rosemarie Emerson, Michelle Mosco, Merle Via.

Last, but not least, numerous people have contributed their time as readers, researchers, providers of photographs, and sources of information and linkages or have made other personal efforts which have been particularly helpful. For this I am grateful to: Ellen Airgood, Nancy Beard, Rob Beekman, Frances Beinecke, Edith and Toby Belcher, Barbara Bielenberg, Peter Borelli, Bob Boyle, Ralph Brill, Bill Burch, Al Butzel, Anne and Frank Cabot, Nash Castro, Janice Corr, John Cronin, Joan Davidson, Nelson Delanoy, John Doyle, Anne Eristoff, Jack Focht, Marc Gerstman, Larry Gobrecht, Mike & Lorenz Graber, The Group, Kathy Hattala, Clay Hiles, John K. Howat, Helen Jimenez, Tom Jorling, Andy Kahnle, Steve Kellert, Bill Kirk, Ronnie and Pascal Knapp, Marcus Koenen, Ann Lapinski, Anthony Lee, Charles Luce, Nancy Mathews, Dave McCoy, Barnabas McHenry, Townley McIlhenny, John Mylod, Raymond O'Brien, Palisades Interstate Park Commission, LtG. Dave Palmer, Roger Panetta, Pierpont Associates, Putnam County Historical Society, Chip and Nora Porter, Jim Ryan, Edward Rutsch, John Sanders, Susan Seligman, Corwin Sharp, Judith Sibley, Tom Siccama, Rosemary Sorkin, Steve Stanne, Peter Stern, Dennis Suszkowski, Ken Toole, Trailside Nature Museum, Dick Wager, Michael Whiteman, Hank Williams.

THE HUDSON RIVER HIGHLANDS

Charles Willson Peale, *After Passing Anthony's Nose, the First Appearance of Sugarloaf Hill.* 1801 watercolor. Courtesy of the American Philosophical Society Library, Philadelphia, Pa.

1. World's End

AT THE SOUTHERN end of Martyr's Reach, navigation charts show a sharp curve in the river marked World's End. The sounding at this point is 216 feet, the deepest spot in the Hudson's 315-mile course from the Adirondack Mountains to the sea.[1] Here on the river bed, under three centuries of accumulated silt, lie the bones of hapless sailors and the rotted hulls of round-bottomed sloops that never arrived at their ports of call.

Four-foot ocean tides wash daily over this cold grave, the calm water surface masking the fast currents and turbulence of a broad river forced through a narrow mountain channel. Whirlwinds and squalls sweep through the hills to confront sailing ships with crosswinds as they round West Point.

It was here, in sight of the northern gate to the Highlands and the placid Newburgh Bay beyond, that Washington Irving's legendary Dolph Heyliger was swept overboard when a sudden storm nearly capsized his sloop. To the south, at a place called Seylmaker's (Sailmaker's) Reach, a real expedition encountered similar difficulties when a delegation of commissioners—which included seventy-year-old Benjamin Franklin—sailed into the Highlands on April 3, 1776. This entry in the journal of Charles Carroll of Maryland describes the conditions he and the other commissioners experienced that day:

> About five o'clock wind breezed up from the south; got under way, and ran with a pretty easy gale as far as the highlands, forty miles from New York. The

> river here is very contracted, and the lands on each side very lofty. When we got into this strait the wind increased, and blew in violent flaws; in doubling one of these steep craggy points we were in danger of running on the rocks; endeavored to double the cape called St. Anthony's nose, but all our efforts proved ineffectual; obliged to return some way back in the straits to seek shelter; in doing this our mainsail was split to pieces by a sudden and most violent blast of wind off the mountains. Came to anchor: blew a perfect storm all night and all day the fourth. Remained all day (the fourth) in Thunder Hill bay, about half a mile below Cape St. Anthony's nose, and a quarter of a mile from Thunder Hill. Our crew were employed all this day in repairing the mainsail.[2]

Maps, journals, and folklore such as these contain clues to the colonists' impressions of the Highlands during the era of exploration and settlement. They tell us that the colonists viewed the area first and foremost as a treacherous passage. Names like World's End, Dunderberg (Thunder Hill), and Seylmaker's Reach appear on the earliest maps of the river as it winds through the narrow mountain pass, warning those who follow of fast currents, winds, and eddies. Where the Hudson flows between Dunderberg and Anthony's Nose, its narrowest point, the current is so strong that sailors called this river reach the Devil's Horse Race, or simply The Race.

The region's reputation as a perilous passage appears to be well deserved. The dangers of navigating the Highlands, where the Hudson narrows from 1 1/2 miles at Newburgh to 3/8 of a mile between Breakneck and Storm King, are well documented. In 1824, the sloop *Neptune*, laden with cargo, sank off Little Stony Point, a section of river known to the Dutch as the Wey-Gat or Wind Gate. "Changing currents, unexpected dead air alternating with gusts" characterize this area according to the *Illustrated Hudson River Pilot*, and "Storm King and Breakneck Mountains funnel the wind currents giving no advantage to either tack."[3]

The record of Henry Hudson's voyage reveals that the crew's journey north through the Highlands was uneventful, but the return trip was not. "Faire weather and the wind at Southeast a stiffe gale betweene the Mountaynes," wrote ship's officer Robert Juet on September 30, 1609. "The high land hath many Points, and a narrow channell, and hath many eddie winds," he noted, as the *Half Moon* lay at anchor to wait for better conditions.[4]

However, danger was not the only concern of the colonial sailors who charted the Highlands. Pollepel Island, for example, was a drop-off place where drunken sailors were left to sober up. In very early maps and journals it appears as "Potlepel Eylant" or Potladle Island, and some historians believe the island takes its name from a popular fifteenth- century Dutch expression "Die polopel hangt eem abt zide" (the potladle hangs from his side), which means, roughly translated, "he's dead drunk."[5]

Buttermilk Falls, Sugarloaf Mountain, and Boterberg, or "Butter Hill," also date to the time when Holland controlled the Hudson Valley. Some names have many spellings, like Popolopen Creek, which shows up on some maps and

The Hudson Highlands and the Hudson River Valley.

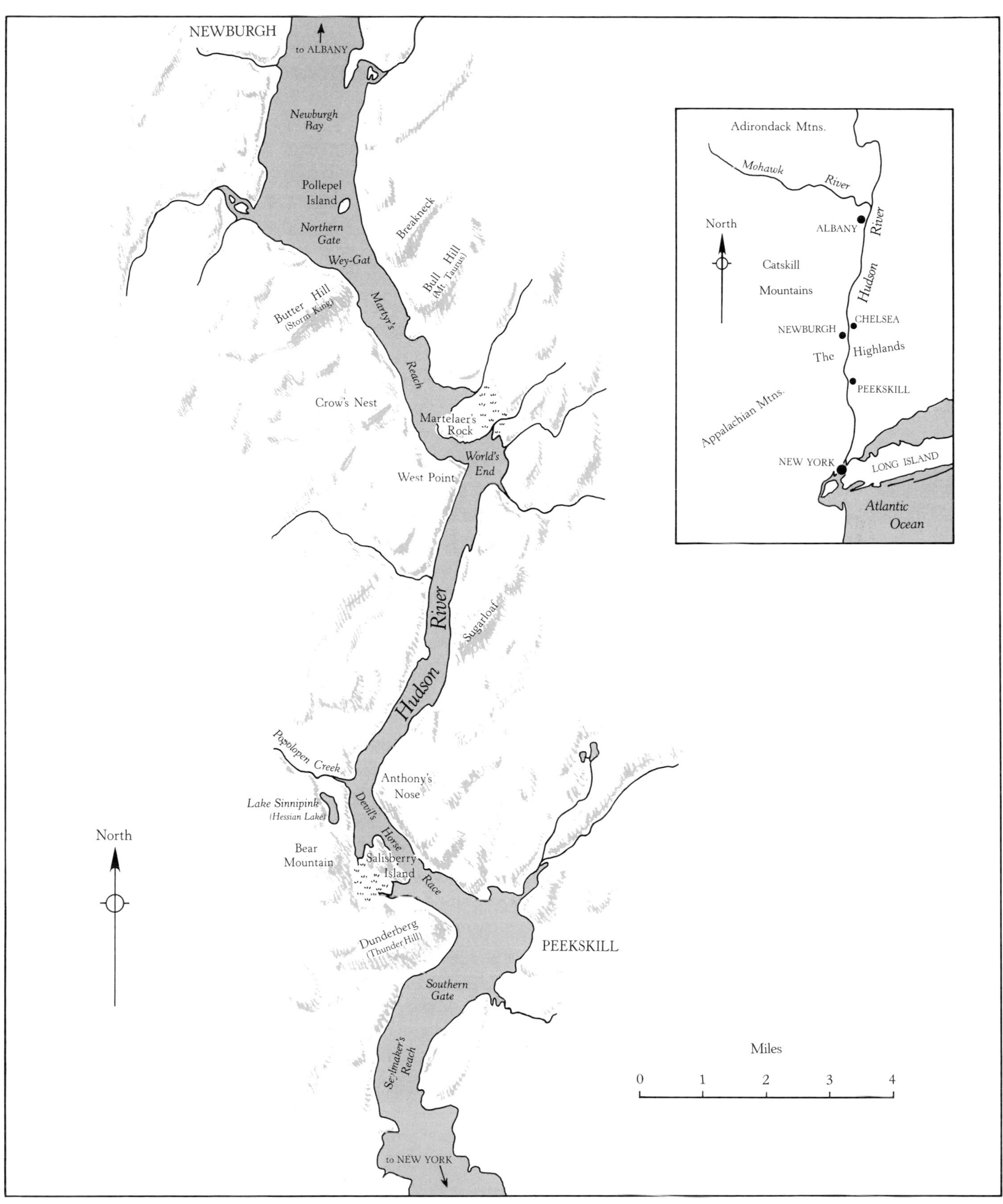
NEWBURGH
to ALBANY
Newburgh Bay
Pollepel Island
Northern Gate
Wey-Gat
Breakneck
Bull Hill
(Mt. Taurus)
Butter Hill
(Storm King)
Martyr's Reach
Crow's Nest
Martelaer's Rock
World's End
West Point
Hudson River
Sugarloaf
Popolopen Creek
Anthony's Nose
Lake Sinnipink
(Hessian Lake)
Devil's Horse Race
Bear Mountain
Salisberry Island
Dunderberg
(Thunder Hill)
PEEKSKILL
Southern Gate
Seylmaker's Reach
to NEW YORK
North
Miles
0 1 2 3 4
Adirondack Mtns.
Mohawk River
North
ALBANY
Hudson River
Catskill Mountains
CHELSEA
NEWBURGH
The Highlands
PEEKSKILL
Appalachian Mtns.
NEW YORK
LONG ISLAND
Atlantic Ocean

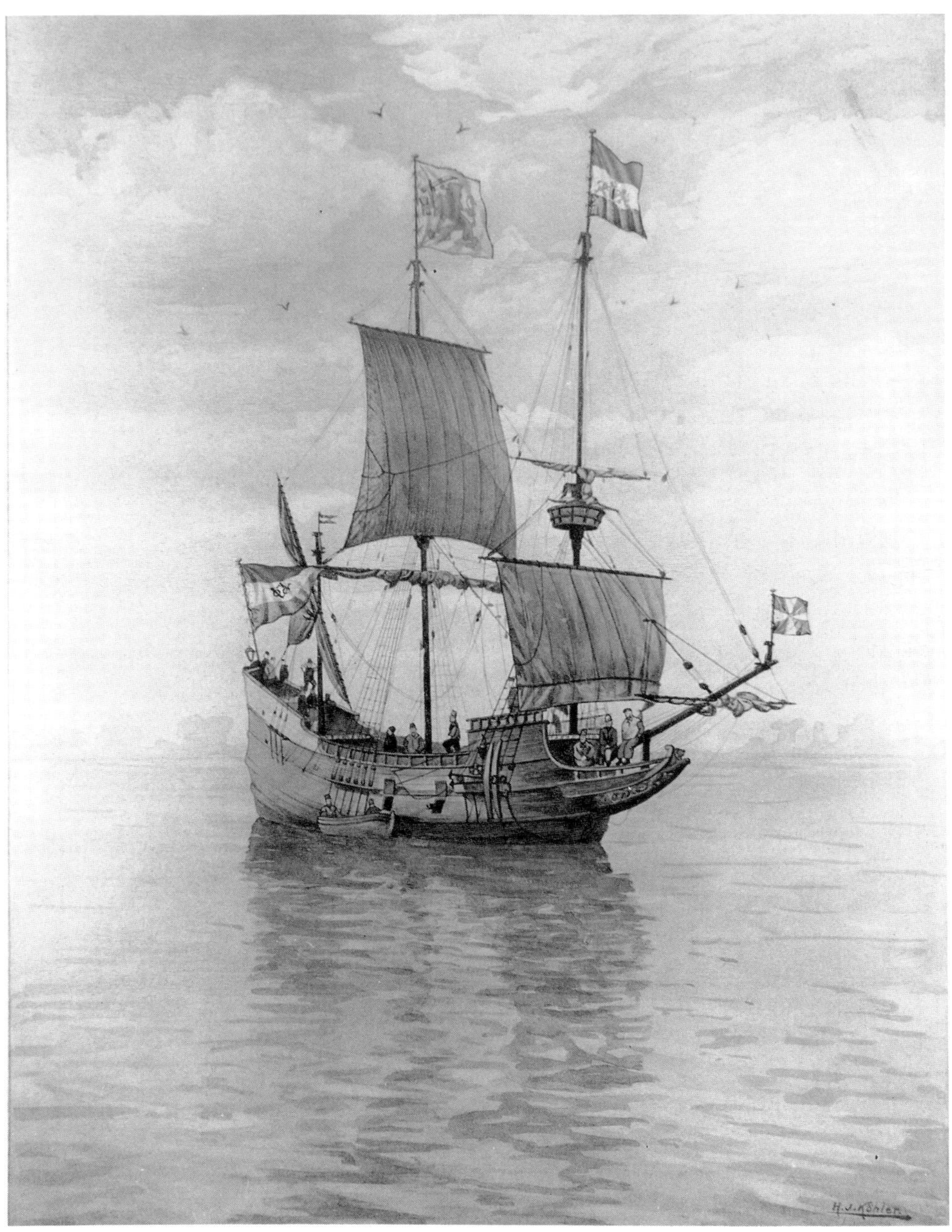

The "Half Moon", showing Henry Hudson's ship published in Amsterdam, Holland for the Hudson-Fulton Celebration, 1909. Courtesy of the New-York Historical Society, New York City.

documents as Pooploop's Kill. Martelaer's Rock was also known as Martyr's and sometimes Marter's Rock. Many of these old place names remain. Notable exceptions are Boterberg and Bull Hill, now known as Storm King and Mount Taurus; and to many people, Pollepel Island is best known as Bannerman's, after its twentieth-century developer.

Some of the names given by the Dutch have an ominous ring, such as Dunderberg, "Thunder Hill," but others seem more humorous. The origin of the name Anthony's Nose (Antonius' Neus) has been a subject of considerable fun and speculation, as this 1836 entry found among the published letters of

Freeman Hunt reveals. It was written after he had visited General Pierre Van Courtlandt in Peekskill.

> General V. is the owner of Antony's Nose, (on the river), as it is called. He gave me the origin of that name. Before the revolution, a vessel was passing up the river, under the command of a Capt. Hogans; when immediately opposite this mountain, the mate looked rather quizzically, first at the mountain, and then at the captain's nose. The captain, by the way, had an enormous nose, which was not unfrequently the subject of good-natured remark; and he at once understood the mate's allusion. "What," says the captain, "does that look like my nose? Call it then if you please Antony's Nose." The story was repeated on shore, and the mountain thenceforward assumed the name, and has thus become an everlasting monument to the memory of the redoubtable Capt. Antony Hogans and his nose.[6]

Folklore provides another insight into the feelings of early settlers towards the Highlands. In Washington Irving's tale of Dolph Heyliger, the hero was forced to swim for his life as his sloop sailed away, carried on by the storm which swept him overboard. On shore a hostile land awaited Dolph. Wild animals, poisonous snakes and thickets of briars and vines confronted his every step. Afraid of being abandoned there, he climbed the nearest hill to search for signs of human life:

> At length he succeeded in scrambling to the summit of a precipice; but it was covered by a dense forest. Wherever he could gain a lookout between the trees, he beheld heights and cliffs, one rising beyond another, until huge mountains overtopped the whole. There were no signs of cultivation; no smoke curling among the trees to indicate a human residence. Everything was wild and solitary.[7]

Danger, practicality, and fantasy are some of the themes that are evident in the early maps, journals, and folklore. The dangers of navigation and the inhospitality of the land are the messages that ring clearest through the colonial age. The early settlers found the Highlands threatening and perhaps even

RATTLESNAKE.

"Wild animals, poisonous snakes and thickets of briars and vines confronted his every step."

supernatural. Judging from folklore, Dutch sailors believed that ghosts and spirits blew up the winds and whirlpools which pulled unlucky sailors into the deep. They felt that primeval forces were at work. To Charles Carroll, whose expedition ran afoul of the weather at Seylmaker's Reach, it was a landscape which could only have been produced by the most powerful earth forces. It was both beautiful and awesome.

> The country round about this bay has a wild romantic appearance; the hills are almost perpendicularly steep, and covered with rocks, and trees of a small size. The hill called St. Anthony's Nose is said to be full of sulphur. I make no doubt this place has experienced some violent convulsion from subterraneous fire: the steepness of the hills, their correspondence, the narrowness of the river, and its depth, all confirm me in this opinion.[8]

From a scientific point of view, Carroll was not far off the mark. The mountain scenery, the serpentine course of the river, and the erratic gusts of wind and weather—all of which combine to create an aura of mystery—are the products of ancient geologic history, which can be read from the exposed rock surfaces and from the living record of trees, plants, fish, and wildlife of the region.

Contemporary studies have shown that the Highland mountains formed as the result of processes which took place deep within the earth. About a billion years ago (during the Precambrian Era) the violent convulsion Carroll speaks of thrust submerged coastal sediments to a depth of ten to twenty miles below ground. There, exposed to the intense heat and pressure of "subterraneous fire," miles-thick layers of sediment turned into rock and in places melted, depositing metals as the fiery liquid cooled.

An examination of the mineral composition and geologic structure of the Highlands reveals that these are durable gneisses and granites. The gneisses, metamorphic rocks which predominate in the Highlands, are streaked with light and dark layers of crystal, indicating that they had once been sediments, later transformed under heat and pressure to hard and resistant new minerals. This bedrock is laced with seams of granite, an igneous rock—the melted form of gneiss, a streaked stone once dissolved into liquid and subsequently resolidified to form a randomly textured material containing crystals of quartz and feldspar. The gneisses and granites are broken and folded. In places the layers and veins appear swirled like marble cake. Concealed beneath the rock surface are lodes of iron ore, deposited as the sediments melted.

This process, which lasted about 400 million years, involved selective melting of liquids and solids. The hot, molten rock moved, and as it reached a cold spot it cooled and solidified in a granite mass; or as cracks formed in the gneiss, liquified rock intruded into the gaps. The evidence of this mingling and intrusion of liquids and solids can be seen where gneiss and granite fade into each other or where veins of black or white igneous rock cut across the finely streaked and layered background of metamorphic gneiss. Selective melting

Highlands gneiss. The dark layers are hornblende, the lighter ones are quartz and feldspar. Photograph copyright © by Jerome Wyckoff reproduced with permission from *Rock Scenery of the Hudson Highlands and Palisades*.

occurred on such a grand scale that some mountains, such as Bear Mountain and Storm King, are composed almost entirely of granite; others, particularly on the east bank, are formed mainly of gneiss.

The gneisses and granites of the Highlands are some of the most ancient rocks in this part of the world—the foundation of the Earth's continents and the basement on which younger rocks have been laid down. Precambrian rocks are exposed in only two places in New York State, the Hudson Highlands and the Adirondacks.

Scientists dispute exactly what happened between the formation of the Precambrian "basement" and its emergence from deep below the Earth's surface to become a mountain chain. The most accepted theory attributes the creation of the Highlands to the movement of giant continental plates which underlie the oceans and mainlands. Geologists believe that two of these independently moving plates crashed against each other in the Precambrian Era. This violent convulsion thrust deep-water strata of the ocean floor over the shallower sediments which had accumulated on the edge of a slowly subsiding coastline, forcing the sediments deep down into the earth, where they became rock and deposited metals. Then, as the plates' impact wore off and the pressure diminished, buoyancy lifted the billion-year-old Precambrian rock toward the surface.

The plates crashed again about 300 million years ago (during the Paleozoic Era), this time with such a slow but powerful force that mountains and valleys were formed along the Atlantic coast in the same way that the hood of a car crumples in a head-on collision. The deep-rooted rocks, now exposed at the earth's surface, were folded, twisted, broken, and lifted thousands of feet high, while earthquakes shook the land. Molten rock invaded the new fractures, leaving crystal veins of granite as it cooled.

This was the first in a series of mountain-building episodes which gave birth to the thousand-mile Appalachian Mountains. From present-day Alabama to Maine, including the Hudson Highlands as well as the Great Smokies, the Blue Ridge, the Ramapos, and the Taconics, the new mountain chain was tall and jagged. It may have been several miles high, like the Himalayas and other younger mountains today. But the weather takes its toll. Today, after the long, slow process of erosion, the mountains have been worn down to their roots.

During the age of dinosaurs, from 200 to 65 million years ago (during the Mesozoic era), the two continental plates rifted apart. These shifts caused the rock mass of the Highlands to rise as a block, while other blocks of the earth's crust sank around it. Once again, the mountains slowly eroded until twenty million years ago (during the Cenozoic Era) when the earth's crust uplifted one more time and raised them again. On the newly elevated land, rivers began to erode valleys. Some of the river valleys coalesced, following fault lines and weaknesses in the basic rock structure, to form the course of the Hudson through the Highlands gorge.

The processes of uplift and erosion over 300 million years ago gave the Highlands their present elevations and softened their contours. However, the landscape was sculpted to its present form primarily by the forces of glaciation.

Beginning in relatively recent geologic history, 2 million years ago, ice caps thousands of feet thick spread south from the subarctic regions of Greenland and Canada, ultimately covering an area of nearly 5 million square miles. The most recent period of glaciation ended only ten to fifteen thousand years ago.

The Ice Age transformed the surface of the land throughout New York and New England. Slow-moving glaciers rounded off ridges and hills, plucked huge boulders from bedrock, and bulldozed tons of soil across vast distances. In the Highlands, the advancing masses of ice sheared off the tip of Anthony's Nose and ground down plains around conical Sugar Loaf, creating hills and valleys. The glaciers also gouged a passage nearly one thousand feet below sea level through the Highlands gorge. Later, retreating glacial ice deposited more than six hundred feet of boulders and rubble in the channel. As the ice melted, the Hudson was filled by the rising sea and the fresh water of its rejuvenated tributary streams.

The passage of the Hudson through the Highlands is one of only two water gaps breaching the Appalachian chain where a river cuts below sea level. The other is Somes Sound in Maine. This makes the Hudson what geologists sometimes refer to as a fjord.[9] Although the river is deep in the Highlands, it is relatively shallow for most of its distance. The federal government maintains a 32 foot channel for shipping, and dredging is periodicly required.

The Hudson is also an estuary, or an arm of the sea, where freshwater and salt water mix. In this respect, it resembles many rivers of the eastern seaboard. It is unusual in that its four-foot tides extend 150 miles upriver to Troy. An Algonquin Indian legend refers to the Hudson as the River-Which-Flows-Two-Ways. According to the legend, the Indians were fated to wander until they found such a river, and on reaching the Hudson Valley, they finally established settlements.

The 150-mile tidal portion of the river flows south at ebb tide, draining the watersheds of the eastern Adirondacks, as well as the Mohawk and the Hudson Valleys. Because the Hudson is more than a mile wide at Newburgh Bay it rushes through the narrow Highlands channel with enormous force. However, at slack tide the current is still, and during flood tide the current actually reverses, moving north. The bottom of the Hudson is below sea level all the way to Troy, and its mean gradient is only 1.5 meters (4.95 feet) over that entire distance. As the Hudson River passes Manhattan's Battery, it flows into upper New York Bay where it is joined by the waters of the East and Harlem Rivers and Newark Bay. It then passes through the Verrazano Narrows into Lower New York Bay, and finally, past Sandy Hook, flows into the ocean.

Tidal action creates an interesting and productive natural environment. Except during spring freshets, salt water mixes with fresh as far north as Chelsea, sixty miles from New York harbor. In drought years, the salinity extends even farther. North of the "salt wedge" the river is tidal but the water is fresh. Alewives, shad, bay anchovies, tom cod, herring, striped bass, and many other fish which spend their adult lives in the ocean require fresh or brackish conditions to spawn. In season, they migrate upriver. Some species, like the shad, travel the full 150 miles of tidal estuary, through brackish water and fresh.

Ocean fish which come into the Hudson to spawn include the prehistoric-looking sturgeon, the striped bass and the shad. The nets of commercial fishermen are shown in the background of this river scene.

Others, like the herring, move beyond the main stem of the river and into its tributaries. For the Atlantic sturgeon, a prehistoric-looking fish that can attain lengths up to eight feet, the deep, cool waters of the Highlands provide ideal spawning habitat. Because sturgeon were once plentiful in the river, early accounts of the Highlands frequently refer to the sounds made by these leaping fish. The leaps, more like giant bellyflops, sent echoes reverberating through the gorge. In the waters at the base of Storm King, the spawning striped bass deposit their eggs.

The natural history of the Highlands was extensively recorded by Dr. Edgar Mearns and published in the *Bulletin of the American Museum of Natural History* in 1898. It has been more recently documented by researchers at Black Rock Forest, an experimental station located behind the crest of Storm King in Cornwall. The ancient granite of the Highlands is thinly covered with rock-strewn soil which supports a forest dominated by oak and hickory. The mountainous topography keeps some areas permanently shaded and cool, while other spots are warmed by the sun. The result is a region capable of supporting isolated outcrops of arctic and subartic species such as black spruce and tundra bog moss, as well as southern plants like the prickly pear cactus.[10] A wide variety of wildlife inhabits the area including bald eagles, which were prevalent in the colonial era and are now returning to the Highlands for overwintering.

Because of the thin, poor soils, the Highlands were one of the last areas of the Hudson Valley to be settled, developing about two hundred years after the colonial manors to the north and south. In 1700, scenery did not figure in the choice of a homestead, and the farmers who immigrated here from Holland, Germany, Scotland, and France chose more fertile lands.

It is the region's unique geology which made the Highlands so inhospitable to colonial explorers. The whirlwinds, changing currents, and unpredictable weather so treacherous for sailors then and now stem from the tidal action and natural features of the land as the Hudson forces its way through the gorge.

P.Lodet, "*The Highlands*," 1806 drawing from the *Hudson River Sketchbook*. Courtesy of the Franklin D. Roosevelt Library, Hyde Park, N.Y.

Because of this, long after the rest of the Valley had been settled, the Highlands were still known primarily as a wild and solitary landscape and a difficult sailing passage to Newburgh Bay and points north. Captains who made it safely through the Race, past World's End, and out of the Northern Gate had put the worst of their voyage behind them. Home once more, they told their children of the spirits which haunted those hills—supernatural beings bent on driving poor sailors to their destruction. It was on just such folklore that Washington Irving later drew for his tale of Dolph Heyliger and the legend of the *Storm Ship*.

The Storm Ship (Excerpt)
by Washington Irving (1832)

> It is certain, nevertheless, that strange things have been seen in these highlands in storms. . . . The captains of the river craft talk of a little bulbous-bottomed Dutch goblin, in trunk-hose and sugar-loafed hat, with a speaking-trumpet in his hand, which they say keeps about the Dunderberg.* They declare that they have heard him, in stormy weather, in the midst of the turmoil, giving orders in Low Dutch for the piping up of a fresh gust of wind, or the rattling off of another thunder-clap. That sometimes he has been seen surrounded by a crew of little imps in broad breeches and short doublets; tumbling head-over-heels in the rack and mist, and playing a thousand gambols in the air; or buzzing like a swarm of flies about Antony's Nose; and that, at such times, the hurry-scurry of the storm was always greatest. One time a sloop, in passing by the Dunderberg, was overtaken by a thunder gust, that came scouring round the mountain, and seemed to burst just over the vessel. Though tight and well ballasted, she laboured dreadfully, and the water came over the gunwale. All the crew were amazed when it was discovered that there was a little white sugar-loaf hat on the mast-head, known at once to be the hat of the Heer of the Dunderberg. Nobody, however, dared to climb to the mast-head and get rid of this terrible hat. The sloop continued laboring and rocking, as if she would have rolled her mast overboard, and seemed in continual danger either of upsetting or of running on shore. In this way she drove quite through the highlands, until she had passed Pollopol's Island, where, it is said, the jurisdiction of the Dunderberg potentate ceases. No sooner had she passed this bourn, than the little hat spun up into the air like a top, whirled up all the clouds into a vortex, and hurried them back to the summit of the Dunderberg; while the sloop righted herself, and sailed on as quietly as if in a mill-pond. Nothing saved her from utter wreck but the fortunate circumstances of having a horse-shoe nailed against the mast, —a wise precaution against evil spirits, since adopted by all the Dutch captains that navigate this haunted river.[11]

*i.e., The "Thunder Mountain", so called for its echoes.

Dutch sloop captains weren't the only ones who understood the difficulty of navigating the Highlands. By 1776, the Continental Congress knew it as well. As a result, World's End, Seylmaker's Reach, and the Horse Race were to figure prominently in the American strategy to avoid British control of the Hudson River during the Revolutionary War.

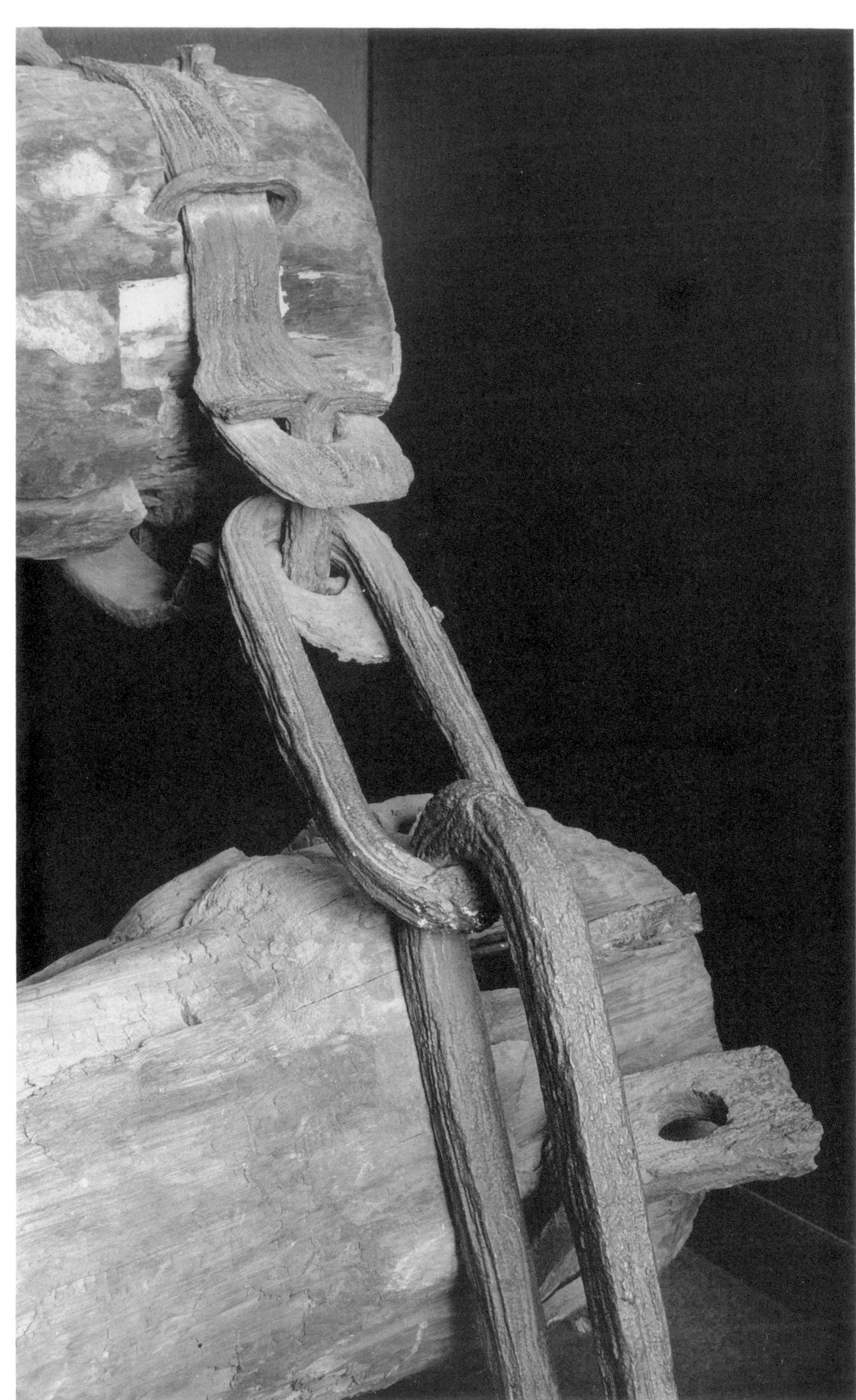

A section of the log boom installed to prevent British ships from sailing up the Hudson River through the Highlands. Photographed at the Washington's Headquarters State Historic Site museum, Newburgh, N.Y. Copyright © 1990 by Robert Beckhard.

2. General Washington's Stronghold

ON JUNE 14, 1776, more than a year since the opening skirmish of the Revolution at Lexington and with the Declaration of Independence soon to be signed, General George Washington wrote to Colonel James Clinton from his New York headquarters:

> Sir: You are to report to Fort Montgomery and take upon you the Command of the Posts in the Highlands. . . . Use every possible Diligence in forwarding the works at Forts Montgomery and Constitution. . . .
>
> As these are or may become Posts of infinite Importance especially the lower one; I cannot sufficiently impress upon you the Necessity of putting them into a fit Posture of Defence, without Delay.[1]

At this time, the theater of war had moved from Massachusetts to New York. George Washington, as commander of the American army, analyzed the moves of his enemy and plotted a defense strategy. The British controlled Canada and the Atlantic coast. Washington concluded that the Royal Navy would soon attempt to take control of the Hudson, thus cutting off all transport between the northern and southern colonies. If they succeeded, the British would have effectively surrounded New England and blocked the movement of supplies and military reinforcements. The Highlands would be the Americans' best choke point to stop the British fleet, and Colonel Clinton, who was very familiar with the area, would be the best man to prepare the forts for attack. Washington's order was none too soon. A month later, in July 1776, General Howe launched the British campaign to quell the American uprising and sent gunboats up the Hudson.

From this time on, the mountain wilderness of the Highlands took on new significance. Military records of the British, the Americans, and their soon-to-be French allies, show that the region became widely known at home and abroad as a site vital to the nation's defense strategy and a stage for heroic dramas of the Revolution. By the time the treaty of peace was signed in 1783, great battles had been fought and lost, espionage and treason attempted, and a new image of the region indelibly stamped on the public consciousness.

New names appeared on military parchments and continued in use after the war was over. Beacon Mountain, Hessian Lake, the Town of Fort Montgomery, Constitution Island, West Point, North Redoubt and Look-out Point all take their names from the revolutionary period. These are not sailor's landmarks but strategic locations on the battleground for national independence.

The events in the Highlands were not the most decisive ones in the Revolution. However, the area was the subject of unwavering attention by both sides throughout the war. The British waged two major campaigns for control of the Hudson, and both focused on the Highlands. In the public mind, the evolving image of the region became that of a citadel, symbolizing the fight for democracy. The stories of these campaigns, which today are long-forgotten, would be told and retold for the next hundred years and fueled major cultural movements of the nineteenth century.

The American decision to fortify the Highlands was reached in 1775, a year before Washington's orders to Colonel Clinton. The *Journals of the Continental Congress* for May 10, 1775 state that a committee had been appointed "to consider what posts are necessary to be occupied in the colony of New York, and by what number of troops it will be proper they should be guarded."[2] Deliberations quickly focused on the Hudson River and the Highlands.

While the patriots were in Philadelphia devising their strategy for New York, the British ministers in London were doing the same. A letter dated July 31, 1775 in London ordered the British forces in America to: "get possession of New York and Albany;—to command the Hudson and East Rivers with a number of small men-of-war, and cutters, stationed in different parts of it, so as to cut off all communication by water between provinces to the northward of it and New York, and between New York and Albany, except for the King's service."[3]

Sir Henry Clinton—a British general thought to be a distant relative of the American Colonel James Clinton—explained the British plan in *The American Rebellion*, a narrative of his campaigns from 1775–1782. He concluded that

> the Hudson naturally presents itself as a very important object. . . . For, as long as a British army held the passes of that noble river and her cruisers swept their coasts, the colonists would have found it almost impossible to have joined or

> fed their respective quotas of troops. And indeed, the inhabitants of the countries on each side must in that case have experienced the greatest distresses, on account of the scarcity of bread corn in those to the east and of black cattle and horses in some of the others, with which they have been accustomed mutually to supply each other. A ready intercourse would also have been by this means obtained with Canada by the lakes (Lake George and Lake Champlain) from whence many obvious advantages might have been derived.[4]

In Philadelphia, the committee of the Continental Congress recommended that the defense of New York rest on fortifications at two locations on the Hudson River—one in Manhattan and one in the Highlands. On Thursday, May 25, 1775, the Congress adopted the following resolution:

> 1. *Resolved*, that a post be immediately taken and fortifyed at or near King's bridge [on the northern tip of Manhattan] in the colony of New York, that the ground be chosen with particular view to prevent the communication between the city of New York and the country from being interrupted by land.
>
> 2. *Resolved*, that a post be also taken in the Highlands on each side of Hudson's River and batteries erected in such manner as will most effectively prevent any vessels passing that may be sent to harass the inhabitants on the borders of said river; and that experienced persons be immediately sent to examine said river in order to discover where it will be most advisable and proper to obstruct the navigation.[5]

The firing distance of cannon, the location of important supply lines, and the capabilities of the British fleet dictated their decision. The Hudson Highlands protected critical ferry crossings between Fishkill and Plum Point and along the King's Road Ferry between Verplanck and Haverstraw. These crossings connected overland routes between New England and the Middle Atlantic colonies. If they were cut, the British could block the movement of supplies and troops.

Located just 50 miles north of New York, the Highlands' forts would also shield the remaining 100 miles of river up to Albany. Furthermore, the mountain range was like a natural fortress, and the narrow, twisting passage of the river through the gorge was ideal for attacking enemy ships. With battlements on both sides of the river, British frigates would be forced to run the gauntlet under a hail of cannon fire. The ascending fleet, slowed considerably by the winds and currrents of the tortuous route and attacked from the bluffs above, would be an easy target.

The Continental Congress transmitted its resolution to the Provincial Congress of New York, which was then meeting in the Exchange Building in New York City. On May 30, 1775, the New York legislators selected two delegates, Colonel James Clinton and Mr. Christopher Tappen, to go to the Highlands and prepare a report on the "proper place for erecting one or more fortifications" and the estimated cost of doing so. On Friday, June 2, Clinton and Tappen hired a sloop and sailed north to the Highlands. Returning a week later, they submitted the following report to the New York Provincial Congress:

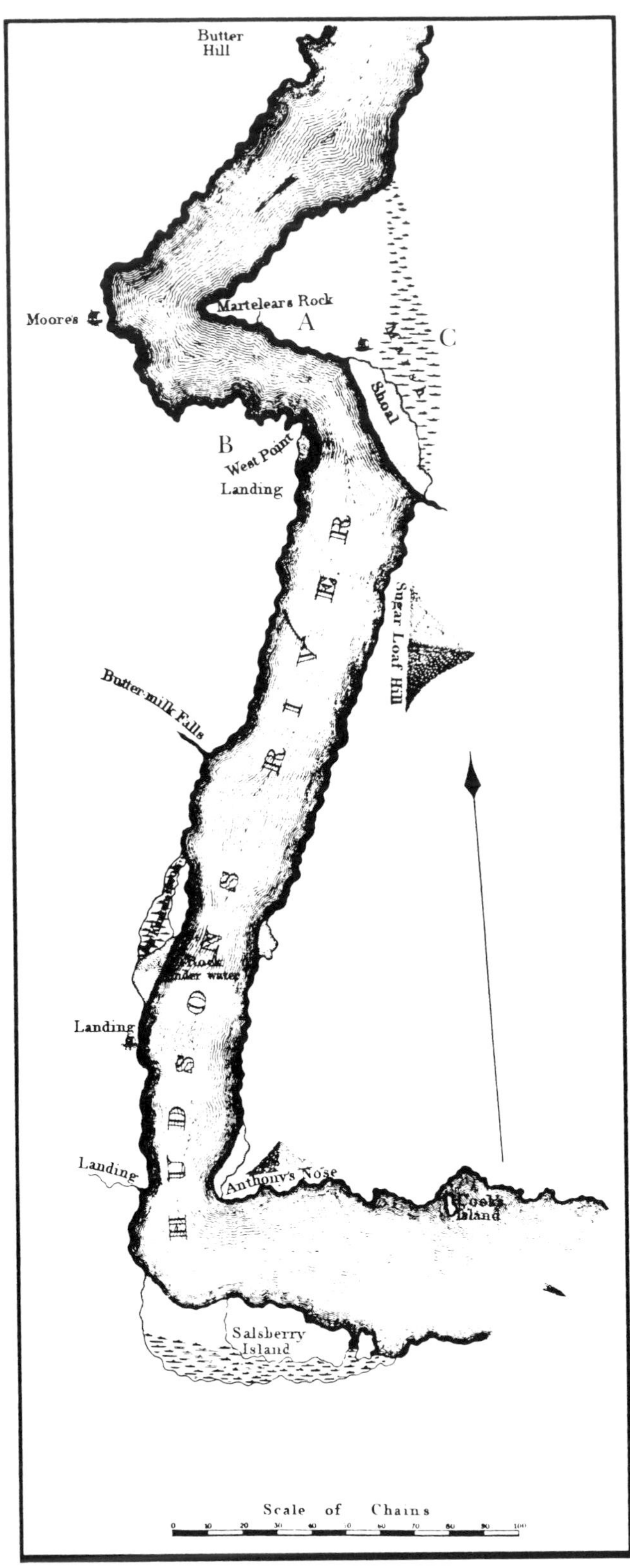

Map showing points to be fortified in the Hudson Highlands drawn to accompany the 1775 report of James Clinton and Christopher Tappen. Courtesy of U.S. Military Academy, West Point.

> Your Commitee . . . report it as their opinion, that a Post, capable of containing three hundred men, erected on the east bank of *Hudson's* River, marked A, and another on the west side of said River, to contain two hundred men, marked B, in the annexed map, will answer the purpose proposed and directed by the Continental Congress, as it is not only the narrowest part of the said river, but best situated, on account of the high hills contiguous to it, as well on the west as east side of the river which cover those parts; so that without a strong easterly wind, or the tide, no vessel can pass it; and the tide on said part of the river is generally so reverse, that a vessel is usually thrown on one side of the river or the other, by means whereof such vessel lay fair and exposed to the places your Committee have fixed on.[6]

The site they had selected was the one sailors refer to as World's End. The committee also recommended that four or five booms chained together could be installed, ready to be drawn across the river to block the passage for vessels.

Work commenced in August 1775 on a fort at Martelaer's Rock. With trained engineers in short supply, the Commissioners hired a Dutch botanist named Bernard Romans to supervise the construction of the works. Romans reconnoitered the terrain and began construction of a "grand bastion" on the southwest end of the island, which the patriots named Fort Constitution, in honor of the struggle for constitutional rights under British law (the American Constitution had, of course, not yet been conceived). Martelaer's Rock soon assumed the name Constitution Island as well. However, Romans ignored the recommendations of the Tappen report for a fort on the west bank opposite Martelaer's Rock and for the construction of a chain and boom to be drawn across the river between the two.

Instead, several months later, Romans and the five New York Commissioners appointed to supervise the fortification of the Highlands began to be interested in another site four miles downriver, just above the reach known as the Horse Race. On October 16, 1775, they submitted a report to the New York Provincial Congress noting that

> at *Pooploop's Kill* (Popolopen Creek), opposite *Anthony's Nose*, it is a very important pass: the river narrow, commanded a great way up and down, full of counter currents, and subject to almost constant fall winds; nor is there any anchorage at all, except close under the works to be erected.[7]

Meanwhile, as investigations continued on sites downstream, progress was very slow at Fort Constitution, over which a bitter dispute erupted between Romans and the New York Commissioners. As a result, on November 8, 1775, Congress sent a delegation of three of its members, Robert Livingston, John Langdon, and Robert Treat Paine, to review the situation. What the Congressmen discovered was that little progress had been made in the four months since construction began. They also urgently recommended that the west side be fortified as well as the east.

"We found the fort in a less defensible situation than we had reason to expect," they reported ". . . and insufficient in itself to answer the purpose of defence." The committee continued with the observation that "the fortress is unfortunately commanded by all the grounds about it, and is much exposed to attack by land; but the most obvious defect is, that the grounds on the *West Point* are higher than the fortress, behind which point an enemy may land without the least danger." They recommended that batteries be erected on the opposite shore, fearing that "nothing short of it will be sufficient to avail us of the winding of the river."[8]

Noting that Romans had identified a place four miles down the river which had higher elevations and less inconvenience, the Congressmen recommended that a delegation be appointed to survey the entire Highlands more thoroughly.

New York then sent a committee of its own to investigate, which reported on December 14, 1775 that "the ground forming the north chop of *Poopkoop's Kill* . . . is by far the most advantageous situation in the Highlands for a fortification," and on January 5, 1776, Congress resolved that "no farther fortifications be erected at Marter's Rock, on Hudson's river, and that a point of land at Poplopen's Kill, on the said river, be without delay effectively fortified."[9] The decision to abandon the original recommendations of the Clinton-Tappen report would be a fateful one.

The following summer, 1776, there was still little progress in fortifying the strongholds in the Highlands. It had been a quiet, cold winter since the British were driven out of Massachusetts. There were attempts at reconciliation, and for months both sides prepared for battle, but as yet the British had not attacked New York. Washington's intelligence sources led him to believe that the British would soon launch an assault. It was at this point that Washington ordered Colonel James Clinton to assume command of the construction. Washington also appointed a new engineer, Thomas Machin, to plan the defenses. Romans had been fired.

With new leadership, progress resumed in June 1776 and American soldiers began building Forts Montgomery and Clinton on the banks of the Popolopen Creek near Bear Mountain. The upper one, Fort Montgomery, was to the north of Popolopen Creek and was named for General Montgomery, who had recently died in the siege of Quebec. The lower one, Fort Clinton, was located on the shoulder of Bear Mountain, to the south of Popolopen Creek, and was named for Colonel Clinton's brother George, Governor of the Province of New York. In August, a third fort, Fort Independence, was begun at the mouth of the Annsville Creek in Peekskill, just south of Anthony's Nose.

Less than a month after Colonel James Clinton assumed command of the Highland forts, General Howe began to probe the lower Hudson defenses. Reinforcements of British troops and German mercenaries had arrived on July 2, and on July 12 two warships, the *Phoenix* and the *Rose*, set sail from New York harbor. To the shock and dismay of American troops, they easily passed the defenses at King's Bridge on Manhattan, carried by a flood tide and a strong wind, and then anchored in the Tappan Zee, twelve miles south of the Highlands. Alarmed, George Washington wrote to General George Clinton on July

Forts and Redoubts of the Hudson River Highlands in the American Revolution.

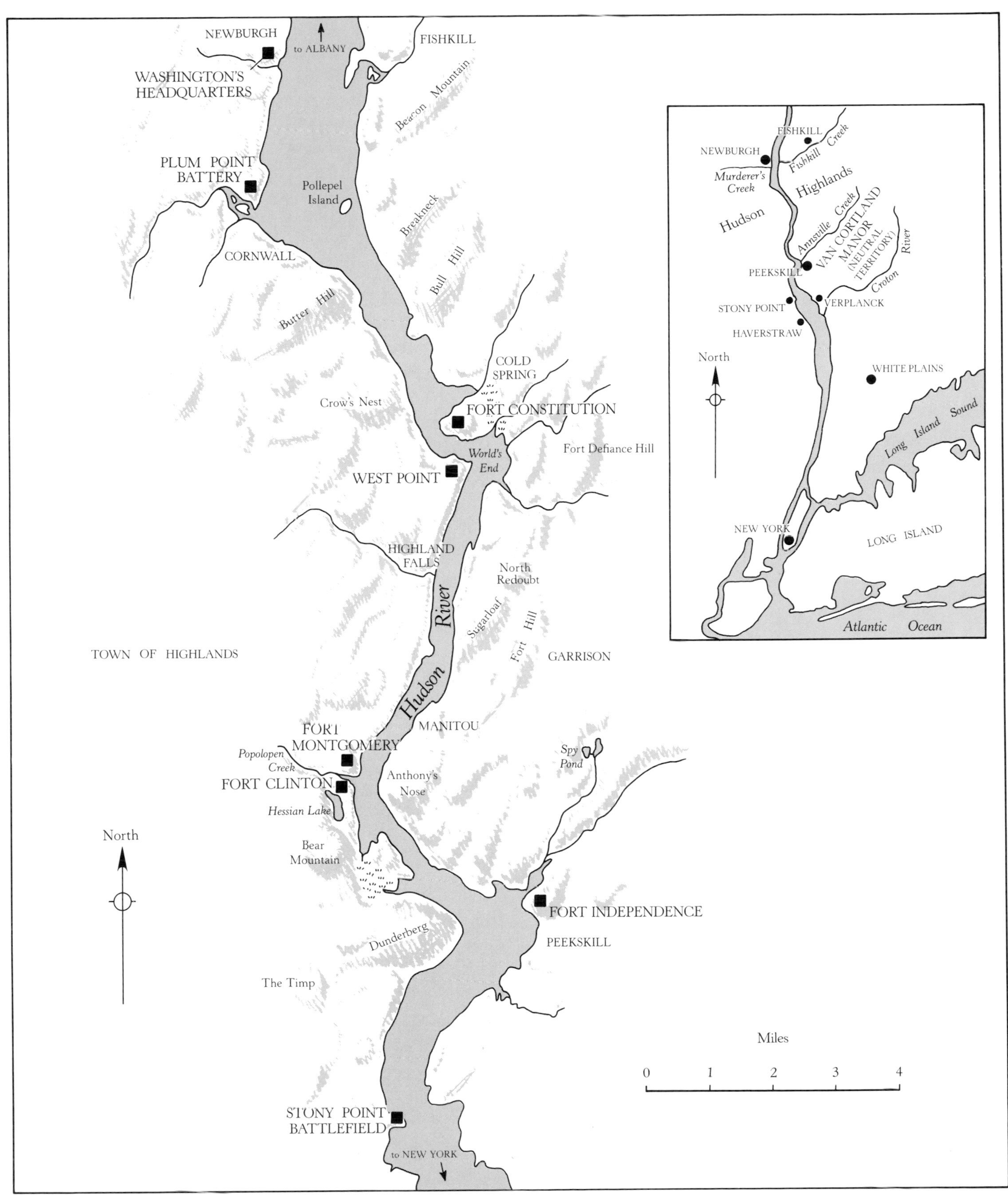
NEWBURGH
to ALBANY
FISHKILL
WASHINGTON'S HEADQUARTERS
Beacon Mountain
PLUM POINT BATTERY
Pollepel Island
CORNWALL
Breakneck
Bull Hill
Butter Hill
COLD SPRING
Crow's Nest
FORT CONSTITUTION
Fort Defiance Hill
World's End
WEST POINT
HIGHLAND FALLS
North Redoubt
Sugarloaf
Fort Hill
Hudson River
TOWN OF HIGHLANDS
GARRISON
MANITOU
FORT MONTGOMERY
Popolopen Creek
FORT CLINTON
Hessian Lake
Anthony's Nose
Spy Pond
North
Bear Mountain
FORT INDEPENDENCE
PEEKSKILL
Dunderberg
The Timp
Miles
0 1 2 3 4
STONY POINT BATTLEFIELD
to NEW YORK
FISHKILL
NEWBURGH
Fishkill Creek
Murderer's Creek
Highlands
Hudson
Annsville Creek
VAN CORTLAND MANOR (NEUTRAL TERRITORY)
River
PEEKSKILL
Croton
STONY POINT
VERPLANCK
HAVERSTRAW
North
WHITE PLAINS
Long Island Sound
NEW YORK
LONG ISLAND
Atlantic Ocean

12: "Sir: Two ships of Force with their Tenders have sailed up Hudson's River. I am apprehensive, that they design to seize the passes in the Highlands."

However, for the time being the British warships stayed put, and a month later, the patriots launched a fireboat attack on the two vessels, successfully burning one of the tenders and sending the gunboats on a rapid retreat to Manhattan. Nevertheless, Washington was shaken by the experience. What particularly concerned him was the ease with which the British warships had passed the American defenses at King's Bridge. In a letter to Colonel Adam Stephen dated July 20, he wrote about the need to supplement forts on land with obstructions in the water:

> Two ships on the 12th. to wit, the Phoenix of 44 Guns and the Rose of 20, exhibited a proof of the incompetency of Batteries to stop a Ships passage with a brisk Wind and strong tide where there are no obstructions in the Water to impede their motion; the above Ships pass'd through an incessant Fire from our Batteries without receiving much damage; they were each hulled several times and their Rigging a little damaged but not so as to retard their way up the River to what is called Taupon [Tappan] bay a wide part of the River out of reach of Cannon shot from either shore.[10]

As a result of this incident, the idea of a chain and boom to be stretched across the river in the Highlands, as outlined in the original Clinton-Tappen report, was revived. One of engineer Machin's tasks was to design the site where the chain and boom were to be attached. Because the main defense of the river in the fall of 1776 now rested on the Popolopen forts, a decision was reached to install the chain below Fort Montgomery instead of at Constitution Island as originally intended. In addition, Machin was instructed to design underwater obstacles betweeen Pollepel Island (now Bannerman's Island) and Plum Point. There the river was shallow enough to construct *chevaux-de-frise*. These were

Chevaux-de-frises, obstructions made of sixty-foot logs, sharpened and tipped with iron points, which were wedged into stone caissons and sunk on the river bottom to gouge the hulls of northbound enemy vessels. Courtesy of the New-York Historical Society, New York City.

obstructions made of sharpened tree trunks tipped with iron points which were wedged into stone caissons and sunk on the river bottom. The chevaux-de-frises protruded just below the surface of the water at an angle designed to gouge the hulls of northbound vessels.

Fortunately for the soldiers building these defenses, there would be no further attacks on the area for more than a year. During this time, Howe engaged and defeated Washington on Long Island, then New York City and White Plains, forcing the Americans to withdraw to a line north of the Croton River. Henceforward, the lands of the Van Cortland Manor, between the Croton River and the Annsville Creek at Peekskill, would be neutral ground between the British and the American territory in the Highlands.

In December 1776, the scene of war shifted to New Jersey. Then, in the summer and fall of 1777, the British mounted a three-point offensive to gain control of the Hudson River. In June, under General Johnny Burgoyne, the red coats marched south from Canada along Lake Champlain. A second force, commanded by Lieutenant Colonel Barry St. Leger, was dispatched from Oswego to capture the Mohawk River. In the fall, Sir Henry Clinton would sail north from Manhattan Island to lead the third prong of the attack, aiming to destroy the rebel forts along the lower Hudson and from there move upriver to meet Burgoyne and St. Leger at Albany.

The majority of the American militia were tenant farmers mustered from the countryside. Beacon fires signaled the call to arms. On every mountain top of the Highlands, five or six men were posted to watch enemy maneuvers. From these redoubts, they would light signal fires which could be seen for miles around. Of the twenty-seven Highland redoubts, Mount Beacon, North Redoubt, and South Redoubt are the best known. On a misty October morning in 1777, smoke rose from the Highland redoubts. Sir Henry Clinton was sailing north.

At the Popolopen forts, the patriot soldiers assembled. They wore farmers' garb. Many of them had no shoes. There were about six hundred of them, black soldiers as well as white. Beneath them, on the river, two frigates, two galleys, and an armed sloop were stationed above the chain to guard the passage. James Clinton and his brother, George, the Governor, assumed command of the two forts and awaited news from the south.

They received word that Sir Henry Clinton had landed an amphibious force at Verplanck's Point on the east shore just south of the Highlands. General Israel Putnam was certain that Sir Henry was aiming for Fort Independence at Peekskill, on the east shore near Anthony's Nose, where he was in charge. The General sent word that he needed reinforcements. Governor Clinton complied, sending troops which had assembled at the Popolopen Forts to join Putnam's troops across the river. Around midnight on October 5, armed British ships sailed from Verplanck to Peekskill Bay with great display of noise and lights.

In the meantime, however, Sir Henry Clinton had quietly assembled two divisions of a thousand men each which he transported at dawn to Stony Point on the west bank, when the river was locked in fog. The attack on Peekskill was

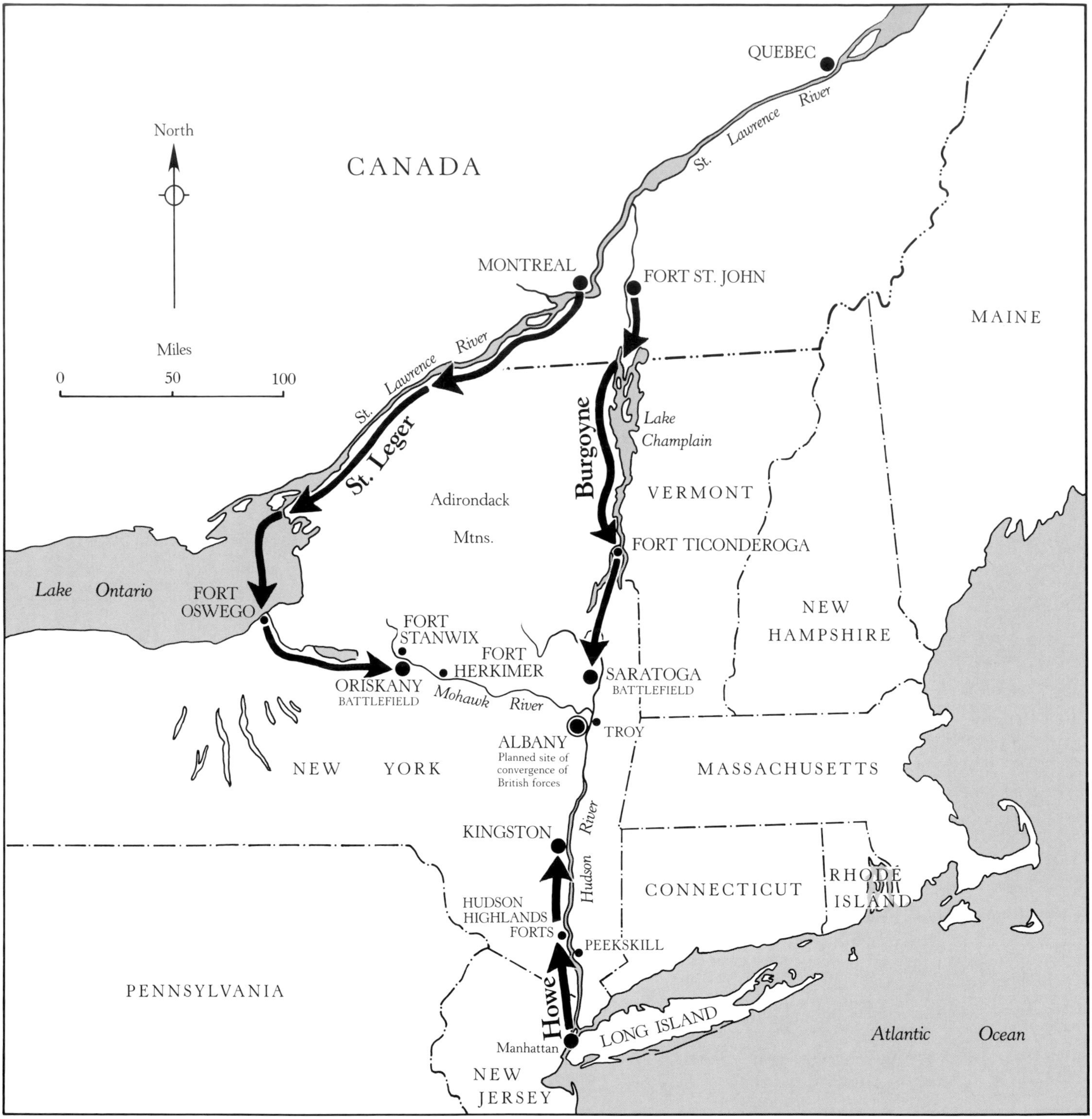

QUEBEC
St. Lawrence River
North
CANADA
MONTREAL
FORT ST. JOHN
MAINE
Miles
0 50 100
St. Lawrence River
St. Leger
Burgoyne
Lake Champlain
Adirondack Mtns.
VERMONT
FORT TICONDEROGA
Lake Ontario
FORT OSWEGO
NEW HAMPSHIRE
FORT STANWIX
FORT HERKIMER
SARATOGA BATTLEFIELD
ORISKANY BATTLEFIELD
Mohawk River
ALBANY
Planned site of convergence of British forces
TROY
NEW YORK
MASSACHUSETTS
Hudson River
KINGSTON
CONNECTICUT
RHODE ISLAND
HUDSON HIGHLANDS FORTS
PEEKSKILL
PENNSYLVANIA
Howe
LONG ISLAND
Manhattan
Atlantic Ocean
NEW JERSEY

merely a diversion, and on the morning of October 6, his troops streamed ashore and marched the twelve miles from Stony Point to the Popolopen forts.

It was a fierce group of soldiers, including Captain Andraes Emerick's Hessian chasseurs and 500 regulars from the 52nd and 57th Regiments, commanded by Colonel Mungo Campbell. They were led through the unfamiliar wilderness terrain by Beverly Robinson and his Loyal Regiment. Robinson had refused to swear allegiance to the rebel cause in 1776 and was forced to leave his house and extensive landholdings in the Highlands at Garrison. He had hunted the area around Bear Mountain since boyhood and had helped Sir Henry Clinton conceive the daring plan. Even so, the terrain was difficult, and it took the better part of the day for the troops to reach their destination. Marching around the back of Dunderberg Mountain through a cut called Timp Pass, at mid-afternoon on October 6, they descended on the forts from behind.

The fighting was bitter and disastrous for the outnumbered and outfoxed American militia. The Highland forts had been designed for firing on ships below. Their flanks were unprotected. By nightfall the British and their hired Hessian soldiers had massacred the patriot defenders. In all, about three hundred Americans were killed, wounded, or captured.

Charles Stedman, who served the British under Sir Henry Clinton on his Hudson River campaigns, gives this 1794 account of the battle:

> In no instance during the American war was more invincible resolution exhibited than in this attack. The British and foreign troops pressed forward silently, under a dreadful fire, and arriving at the foot of the work, actually pushed one another up into the embrasures. The garrison, consisting of four-hundred men, for a little while longer contested the rampart. Some of our men were killed in the very embrasures, and several were wounded with bayonets in the struggle; so that it must be admitted the Americans defended themselves courageously. At length the rampart was cleared. The Americans retiring to the other side of the esplanade, discharged a last volley, and threw down their arms.[11]

Leading the British rout of one of the two forts was Beverly Robinson, who assumed command after Colonel Mungo Campbell had been shot. Robinson was the first man over the ramparts, and some of the patriots killed in battle had been tenants on his own farms across the river. The forts were virtually defenseless, especially from the rear, with walls not more than half raised in places. Nearly all the garrison fled, among them James Clinton, who hurried to assist his brother at the other fort. There, seeing that all was lost, he was able to arrange cover for the Governor's escape and himself managed to flee that night.

Realizing that their ships were now vulnerable to attack by the British from the forts above, the patriots guarding the chain slipped their cables and attempted to sail up river. However, the same conditions which made the Highlands difficult for advancing British frigates plagued the American vessels; and with adverse winds preventing their escape, the Americans set fire to their frigates and let the burning vessels float downstream toward the enemy.

Map of the theater of war in New York in the Revolutionary war.

Even in the pitch of battle, the powerful effect of the mountain scene was not lost on the British officer, Charles Stedman, who later wrote of it:

> The flames suddenly broke forth; and, as every sail was set, the vessels soon became magnificent pyramids of fire. The reflection on the steep face of the opposite mountain [Anthony's Nose], and the long train of ruddy light which shone upon the water for a prodigious distance, had a wonderful effect; whilst the ear was awfully filled with the continued echoes from the rocky shores, as the flames gradually reached the cannon. The whole was sublimely terminated by the explosions, which left all again to darkness.[12]

Two days later, on October 8, 1777, the British armada under General Vaughn once more set sail and took Fort Constitution without a fight. Their ships easily passed the underwater obstacles at Pollepel Island, continuing north to Kingston, the provincial capital, which they burned on October 16.

Despite the loss of the Highland forts, the British attempt to control the Hudson was foiled. Burgoyne was defeated at Saratoga on October 17, before Sir Henry Clinton could reach him with reinforcements. St. Leger's forces failed at Oriskany, where he was stopped by Benedict Arnold's heroism. Sir Henry Clinton could not hold the Hudson alone without risking the loss of Manhattan, which needed his troops for protection. The British fleet held the Highlands for twenty days before it was forced to withdraw. In a final act of defiance, American troops stationed at Plum Point to guard the chevaux-de-frise opened fire with five cannon and raked the British armada as it retreated through the Highlands to New York.

In the spring of the following year, 1778, Timothy Dwight, a young chaplain in the American army, visited the area. His observations, published later when he was president of Yale College, provide a record of the ravages of war in the Highlands.

> I went down the river in company with several officers to examine the Forts Montgomery and Clinton, built on a point six or eight miles below West Point for the defense of the river. . . . We found, at a small distance from Fort Montgomery, a pond of moderate size in which we saw the bodies of several men who had been killed in the assault upon the fort. They were thrown into this pond, the preceding autumn, by the British, when probably the water was sufficiently deep to cover them. Some of them were covered at this time, but at a depth so small as to leave them distinctly visible. Others had an arm, a leg, and a part of the body above the surface. The clothes which they wore when they were killed were still on them, and proved that they were militia, being the ordinary dress of farmers. Their faces were bloated and monstrous, and their postures were uncouth, distorted, and in the highest degree afflictive. My companions had been accustomed to the horrors of war, and sustained the prospect with some degree of firmness. To me, a novice in scenes of this nature, it was overwhelming.

After the war, Lake Sinnipink, behind Fort Clinton on Bear Mountain, was known as Bloody Pond, because so many of the dead were cast into its waters. It is now called Hessian Lake. Although Dwight's description of this scene is a gruesome one, his letters for the same period show an appreciation of the area's beauty and indicate that its wilderness character was no longer as frightening as it had been a century earlier:

> Yesterday afternoon, in company with Major Humphreys I went up to the summit of Sugarloaf, a mountain near Col. Robinson's house. . . . Everything which we beheld was majestic, solemn, wild, and melancholy. . . . Directly north, the Hudson, here a mile in breadth, and twice as wide higher up, is seen descending from a great distance, and making its way between the magnificent cliffs of the two great mountains, Butter Hill and Breakneck. The grandeur of this scene defies description. Through the opening, here called the *Wey-gat or Wind-gate*, because the wind often blows through it with great violence, is visible the cultivated country at New Windsor throughout a considerable extent. Beyond this, at the distance of about forty miles, rise the Catskill Mountains, whose blue summits were at this time lost in the clouds. In this reach of the river lies an island, to the eye a mere bird's nest, and near it were two boats, resembling in size those which children make of paper.[13]

By November 1777, the British retreat had returned the Highlands to American control. Burgoyne's defeat at Saratoga was a turning point in the war. The British abandoned the idea of using the Hudson-Champlain route, and within a few months, the scene of war shifted to the south. News of the surrender of 5000 British and German troops reached France, where Louis XVI was considering an alliance with the new republic. The momentum of the Revolution was building. France signed a treaty pledging full military support and was followed soon after by Spain and the Netherlands. The war became a global one, and the British drew off many of their troops from New York to defend the previously unthreatened West Indies.

That these events would soon unfold was not known to George Washington in the fall of 1777. With the downfall of Forts Montgomery and Clinton at the southern gate to the Highlands, he immediately decided to relocate the American defenses to West Point, opposite Constitution Island, as originally recommended in the Clinton-Tappen report.

Construction began in January 1778, and a new engineer, Colonel Louis La Radière, a Frenchman, began to outline a new fort on the plateau at West Point. However, he was replaced a few months later by Colonel Thaddeus Kosciusko, and it was under him that the fortifications began to take shape. The massive undertaking soon won the Polish colonel great admiration. Engineer Thomas Machin remained and concentrated on designs for a new chain.

The plan for West Point called for a fortified area consisting of mutually supporting strongholds. The Marquis de Chastellux, who visited West Point in 1780, while he served the French allied forces, commented on the "very intelligent manner in which they are calculated. . . .

"From the fort of West Point properly so-called, which is on the edge of the river, to the top of the mountain at the foot of which it stands, are six different forts, all in the form of an amphitheater, and protecting each other."[14] Being

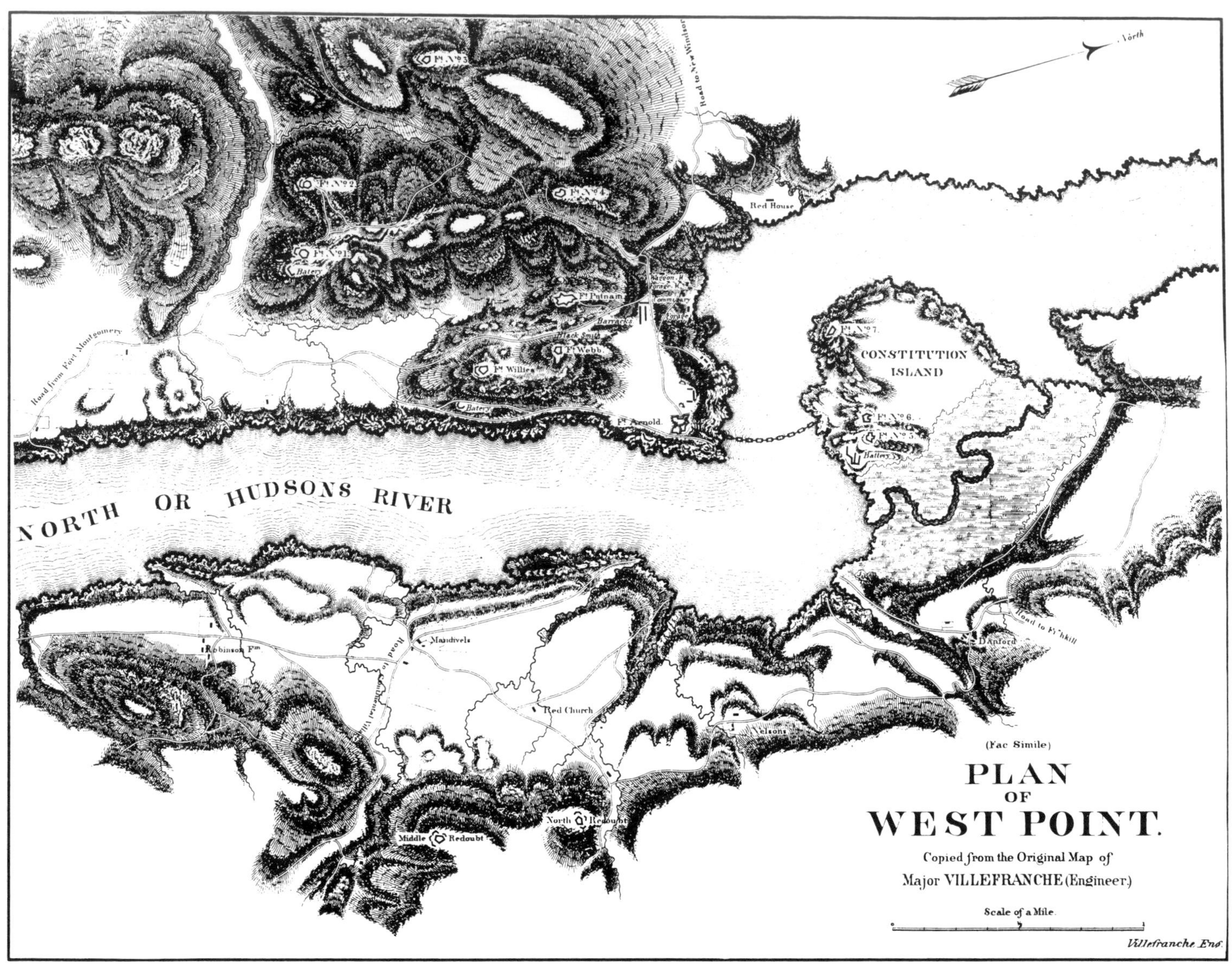

Major Jean Louis Ambroise de Genton Villefranche, engineer, *Plan of West Point*, facsimile engraving from the original 1780 map. From a copy in the Special Collection Division, U.S. Military Academy Library, West Point.

much easier to defend than the traditional "single position" of the eighteenth century, this system of fortification was far ahead of its time. The main garrison was constructed on the plain, a flat area on the bluffs above the river. It was named Fort Arnold in honor of General Benedict Arnold, still the hero who had recently routed St. Leger at Oriskany and Burgoyne at Saratoga. Above the plain, Fort Putnam defended Fort Arnold and the lower river positions. It also covered three smaller forts on the ridgeline to the south. Approaches to Fort Putnam were protected by redoubts 1–3 while redoubt 4 protected the ground above Fort Putnam. Battery 1, constructed below redoubt 1, provided a position from which to fire on enemy ships.

On Constitution Island, two low-lying batteries covered the river and a new chain, while redoubts constructed along the crest of the island protected the

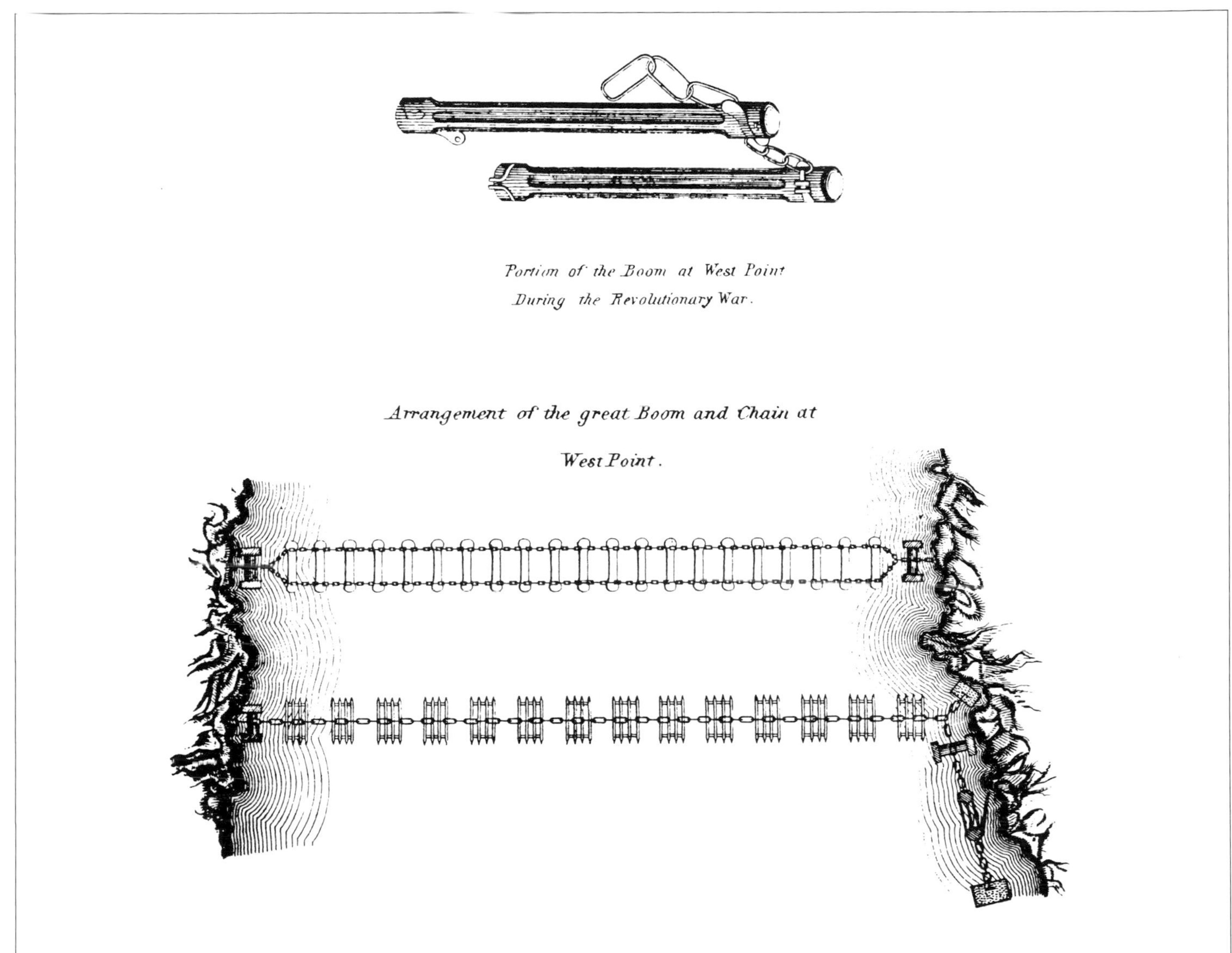

Portion of the Boom at West Point
During the Revolutionary War.
Arrangement of the great Boom and Chain at
West Point.

The chain and boom at West Point. Courtesy of U.S. Military Academy, West Point.

river batteries from landward attack. A new iron chain blockaded the river between West Point and Constitution Island. This second chain, of the finest workmanship and forged with the highest quality ore, was considerably stronger than the one broken by the British at Fort Montgomery a year earlier. The Sterling Furnace near Harriman burned night and day for four months, producing the links in groups of ten. These groups were joined together at nearby New Windsor and then floated into place just upstream of World's End. Each link was two feet long and weighed 180 pounds. The chain was secured below to a huge wooden boom intended to form an impenetrable barrier to the British fleet.

By 1780, the scene of war again shifted, and both British and American forces began to contemplate moves in New York. After a year of disasters, the promise of French reinforcements under Count Rochambeau led George Washington to conclude that he might have sufficient force to drive the British from New York. The man Washington had in mind to command the left wing of the army for an attack on Manhattan was Benedict Arnold.

When Washington met with Arnold to offer him command of the left wing, Arnold was sullen and strangely quiet. "His countenance changed and he appeared to be quite fallen," recalled Washington, "and instead of thanking me . . . never opened his mouth."[15] He later returned and told Washington that he wished to be given command of West Point instead. Washington did not hesitate to grant the request, but he was mystified that Arnold should turn down combat assignment for an assignment not nearly as important.

Arnold was one of the Americans' most trusted and courageous generals. At Quebec, Valcour Island and Danbury, Arnold had demonstrated superb combat leadership. In May 1777, at Saratoga, he was so fearless that some suggested he had been momentarily insane, or had taken opium, or that he simply wanted to die. His daring attacks on the British lines had inspired an American victory. The choice of a non-combat position seemed quite uncharacteristic. However, Arnold claimed his injuries at Saratoga were too debilitating, and Washington complied with his request. This request was Arnold's first act of treason, and it set in motion the British plot to capture West Point.

No one knows what really caused Arnold to turn traitor, but it is clear that he had felt a growing disaffection since his earliest actions in the war. Despite his distinguished service and repeated heroism on the field of battle, he was consistently passed over for promotions and desired positions of command.

In 1778 Arnold was assigned to Philadelphia, where, in 1779, he was censured by a court martial for poor administration of a command. This appears to have been the final blow to his ego. It was soon afterward, in Philadelphia, that Arnold contacted an aide to the British General, Sir Henry Clinton, and offered, for a price, to provide his services to the crown. In July 1780, after a year and a half of negotiating, Arnold informed Clinton that he would receive command of West Point. The British general agreed to Arnold's offer of treason.

They settled on a fee of 20,000 pounds sterling if Arnold was successful. This small fortune would allow Arnold to support his new wife, teenage Peggy Shippen, a socialite and Tory sympathizer, in lavish style. Clinton also made it clear that Arnold would be paid if he failed, although it would be a lesser sum. It was enough to convince Arnold to cast off his allegiance to the rebel cause. From that time on he began plotting the British capture of the Hudson River, focusing on the Highland forts. Arnold took command of West Point in August 1780, and a month later, in a secret meeting with Major André of the British Army, he passed over the plans of the fortification.

Arnold and Major André met on the banks of the Hudson near Haverstraw at midnight on Thursday September 21, 1780. André rowed over from the British ship *Vulture*, with a pass signed by Arnold, and discussed West Point with its new commander until dawn. As André prepared to return to his ship, he heard the sound of cannon fire on the river. American troops had shelled the

Vulture, and the ship was forced to retreat downriver. This left André stranded, with no recourse but to travel the risky overland route through the "no man's land" between the American strongholds in the Highlands and the British lines at the Croton River.

With Arnold's help, André disguised himself as a civilian, and, carrying the plans to West Point in his boot, he crossed the Hudson by boat to Verplanck's Point and began the thirty-mile ride to the British headquarters at White Plains. André was accompanied for the first fifteen miles by an American, Joshua Hett Smith. The two men carried passes signed by Arnold, which they used when stopped by a detachment of militia near Crompond, south of Verplanck's Point. Spending the night at Crompond, they continued on another seven or eight miles, when Smith informed André that he would have to continue the rest of the way alone.

Approaching Tarrytown and safety, André was accosted by three volunteer militiamen. Mistaking them for Tories, André declared that he was British. The three Americans then searched him and pulled his boots off, discovering the maps and plans of West Point. André sought to bribe the militiamen into letting him continue, but this merely convinced them that he had something of value, and they turned André in to the American command. The plot to betray West Point had failed.

Arnold was at his headquarters in Garrison when he received word of André's capture. He learned that a letter implicating him was underway to George Washington, who at that very moment was on his way to the Highlands. Arnold went upstairs to tell his wife the news and then hastily departed, announcing that he had to go across the river to prepare for Washington's arrival at West Point. Spurring his horse to a gallop, he raced to Beverly Dock, where a boat was waiting to carry him across to West Point. Commanding the confused but obedient boatmen to ferry him instead to the British ship *Vulture*, which had

No. 5.

[In the handwriting of Major Bauman.]

RETURN OF THE ORDNANCE IN THE DIFFERENT FORTS, BATTERIES, &c.. AT WEST POINT AND ITS DEPENDENCIES,
Sept. 5, 1780.

	Metal.	Garrison Carriages.	Garrison Carriages.	Traveling Carriages.	Garrison Carriages.	Stocked Carriages.	Garrison Carriages.	Garrison Carriages.	Stocked Carriages.	Traveling Carriages.	Garrison Carriages.	Traveling Carriages.	Traveling Carriages.	Mortars. Inches.	Mortars. Inches.	Mortars. Inches.	Howitzers.	Total.
Calibers		24	18		12		9	6		..	4		3	0	5¼	4⅖	8	
Fort Arnold	Brass.											1		5	5	1		23
	Iron.	1	6			1							3					
Fort Putnam	Brass.									2		1			4			14
	Iron.		5		2													
Constitution Island	Iron.				4		1	5										10
South Battery	Iron.		4		1													5
Chain Battery	Iron.				1	2												8
Lanthorn Battery	Iron.						2											2
Webb's Redoubt	Iron.				1				2		1							4
Sherman's Redoubt	Iron.							2	3									5
Megg's Redoubt	Iron.							1				1						2
South Redoubt	Iron.				1			4										5
North Redoubt	Iron.		3		3													6
Wyllys's Redoubt	Iron.			2									3					5
Rocky Hill, No. 4	Iron.							2										2
" No. 1	Iron.				1		4											5
" No. 2	Iron.						2											2
Verplank's Point	Brass.														2	1		8
Stony Point	Iron.			1		2											1	4
Total		1	18	8	14	5	9	14	5	2	1	8	6	5	11	2	1	100

N. B. The following ordnance not distributed:

No. 6 Iron 12 Pounder.
4 do. 9 do.
1 do. 6 do.
1 do. 4 do.
2 do. 8 do.

14

8 Brass 24 Pounders.
7 do. 12 do.
1 do. 8 inch howitzer.

11

S BAUMAN,

Major Com'd of Artillery.

Among the papers provided by Benedict Arnold to Major Andre was this listing of ordnance at West Point, (transcribed from the handwriting of Major S. Bauman) Courtesy of the U. S. Military Academy, West Point.

retreated to the south, he escaped. Arnold was given a pension in the British army; André was tried and hanged.

News of the event spread like wildfire. On October 10, the *Pennsylvania Packet* printed General Greene's orders of the day for September 26:

> Treason of the blackest dye was yesterday discovered. General Arnold, who commanded at West Point, lost to every sentiment of honor, of public and private obligation, was about to deliver up that important fort into the hands of the enemy. Such an event must have given the American cause a deadly wound if not a fatal stab.[16]

The people of Philadelphia were so outraged that they carted an effigy of Arnold through the streets to be hung, and then burned it. Later, George Washington was to write: "That overruling Providence which has so often, and so remarkably interposed in our favor, never manifested itself more conspicuously than in the timely discovery of his horrid design of surrendering the Post and Garrison of West Point into the hands of the enemy."[17]

As part of Arnold's plan, a link of the iron chain across the river was to have

The Hasbrouck House in Newburgh, Washington's Headquarter's from 1782–1783, now preserved as a state Historic Site. From Lossing's *Life of Washington*.

A. Van Zandt, *The Hudson North, from West Point,* undated wash on paper, showing a peacetime view of the Highlands. Courtesy of the Hudson River Museum, Yonkers, N.Y.

been removed for "repairs," but André's capture prevented the British from putting this scheme to the test. The chain remained in place for the duration of the war. Close to one hundred years later, during the Civil War, most of it was melted down for reuse. A few links were saved and may be seen today at West Point.

With André's capture, Sir Henry Clinton abandoned any thought of taking West Point. The remaining battles of the war were fought in the south; and in 1781, the British, under Cornwallis, suffered a crushing defeat at Yorktown, signaling the end of the Revolution.

In 1782, Washington moved his headquarters to the Hasbrouck House in Newburgh, just north of the Highlands, and remained there for the duration of the war, until the cessation of hostilities was announced and ratified by Congress in 1783. For George Washington, the disadvantages of an untrained militia in warfare would not quickly be forgotten. No sooner was he elected President than he addressed Congress on the need to establish a military school. In 1802, after his death, the lawmakers funded a military academy to be located at West Point.

The seven years of the Revolution, though brief in the history of the Highlands, left a lasting impression in the public mind. A new image was taking hold, a "wild and warlike abode" evoking the "dominion of the God Mars" as the Marquis de Chastellux described the Highlands in 1780.[18] Years later, West Point would be referred to as America's Gibraltar, and by the turn of the century, merely the mention of the Highlands was sufficient to recall thoughts of George Washington, American heroism, and the birth of a democratic nation. Britisher John Maude, expressed these thoughts in his journal of a sloop voyage up the Hudson River in 1800:

"Turned out at 4 a.m. Sketched a view of Fort Clinton, Fort Montgomery, St. Anthony's Nose, the Bear Mountains, and surrounding scenery; highly romantic and beautiful being the entrance to the Highlands. . . . The view from our present situation (West Point) was most sublime and magnificent. I do not recollect one that I enjoyed so much; it was historic ground and had been trodden by Washington."[19]

3. West Point: First Stop on the American Tour

THE INVENTION of the steamboat in 1807 brought great changes to the Hudson Valley, introducing an age of leisurely travel. This was a time when sightseeing was coming into international popularity and coincided with a period of intense nationalistic sentiment among Americans based on a fierce pride in the nation's history and scenery. What better way to see America than from aboard a paddlewheeler, and where better to feel a surge of patriotism than floating by the Highlands' historic forts and romantic landscape? Of course, this was not the only place in the country where spectacular scenery could be viewed, nor was it the only location of Revolutionary War interest. However, in the Highlands both history and scenery could be enjoyed in one place. Furthermore, the location of the region, a mere fifty miles from New York (the port of entry for foreigners) on a river well served by steamboats, made it accessible to Europeans and city dwellers alike, and the Highlands shot into national and international prominence. The new military academy at West Point, established by Congress in 1802, with its awe-inspiring views and its grounds "trodden by Washington," would be a big draw for visitors from home and abroad for decades to come, all made possible by Robert Fulton's invention.

The *Clermont*'s maiden voyage on August 17, 1807 was a historic event. The vessel, which was described as "an ungainly craft looking precisely like a backwoods saw-mill mounted on a scow and set on fire," traveled from New

York to Albany and back, a distance of 300 miles, in only 62 hours. The same round-trip passage would have taken a sloop about a week, depending on the tides and weather. "I overtook many sloops and schooners, beating to the windward, and parted with them as if they had been at anchor," wrote inventor Robert Fulton of the voyage.[1]

Steamboats rapidly won acceptance as comfortable, fast and affordable ways to travel. In 1822, New York State Governor DeWitt Clinton advertised that one could travel two hundred miles in four days without a jolt or fatigue for the small sum of eight dollars. Steamboats had the advantage over sloop or carriage transportation of allowing plenty of time for conversation, amusement or "viewing the most interesting region of the globe."[2]

By 1850, after the Fulton monopoly was broken, there were nearly 150 of these vessels plying the Hudson, carrying as many as a million passengers annually, many of them Europeans who flocked to the new republic's shores in increasing numbers. These "floating palaces" measured as much as 350 feet long and could carry several thousand people. Captain Frederick Marryat, a British author of popular sea-faring novels, decribed the impression they made on him:

A.E. Emslie, *"Up the Hudson,"* July 30, 1870 engraving from *Harper's Weekly.* Courtesy of Steve Dunwell.

Varick de Witt, *Steamboat "Clermont" on North River,* 1861 watercolor, showing Fulton's invention as it may have appeared on its maiden voyage as it passed the Palisades, south of the Highlands. Courtesy of the New-York Historical Society, New York City.

"When I first saw one of the largest sweep round the battery, with her two decks, the upper one screened with snow-white awnings—the gay dresses of the ladies—the variety of colours—it reminded me of a floating garden."[3]

The excitement of traveling by means of Fulton's novel invention was matched only by the passion for viewing the Hudson's river scenery. Frenchman Jacques Milbert was amazed at the "national enthusiasm for these magnificent scenes." His compatriot Edouard de Montule reported that the conversation of his traveling companions, "like that of most Americans, had always for its subject the excellence of their country. . . . Although this national spirit, which is nurtured by the newspapers, is a near relative of conceit, it is very beneficial for a people. . . . the Romans, too, had this pride; it made them and kept them for a long time the masters of the world."[4] This fascination with scenery held the attention of Americans and Europeans alike for at least fifty years and lingered as long as steamboat travel was popular, into the twentieth century.

The early nineteenth century saw America seeking to define its own distinctive culture. With fifty years of independence behind the nation, and the War of 1812 over at last, a sense of national identity was slowly emerging, but it lacked shape and substance. People needed something to hang their hats on. Westward expansion and immigration had changed the character of the country. Culturally, it was drifting. In the fields of art, literature, and architecture, the new nation had little to show for itself. Its history and traditions, when judged against a European standard, could hardly compare to the cultural heritage of Greece and Rome. However, in one area it was clear that America was incomparable. Its wilderness had no counterpart in Europe. To be sure, there were patches of wild land in Europe, but America was a wild continent.

By this time, wilderness was becoming fascinating—no longer so frightening as it had been a century earlier—and people began to look at the land in a new way. Governor DeWitt Clinton was one of the first to challenge his countrymen to take pride in the nation's resources and, in fact, to make American wilderness and scenery a wellspring of its culture. In an 1816 speech before the National Academy of Fine Arts, he asked:

> Can there be a country in the world better calculated than ours to exercise and exalt the imagination. . . ? Here nature has conducted her operations on a magnificent scale. . . . This wild, romantic and awful scenery is calculated to . . . exalt all feelings of the heart.[5]

In addition to the nationalists' scenic fervor was a glowing pride in the recent victory of independence, an accomplishment which was equally admired abroad. The American Revolution offered proof of virtue and a symbol of the triumph of democracy. Stories and scenes of the war, such as Arnold's treason and the storming of Forts Montgomery and Clinton, were cherished.

In an age of steamboat travel, the Highlands and West Point soon attracted international attention as a place where the finest qualities of American heritage and scenery could be enjoyed. Charles Dickens, one of many foreign visitors, expressed a commonly held sentiment when he wrote in the conclusion of his book *American Notes*:

> Among the fair and lovely Highlands of the North River: shut in by deep green heights and ruined forts . . . hemmed in, besides, all around with memories of Washington, is the Military School of America. It could not stand on more appropriate ground, and any ground more beautiful can hardly be.[6]

The effects on the Highlands were far-reaching. This small, mountain landscape became the focus of a quest for national identity. Its ruined forts assumed the importance of Grecian temples. The stories and scenes of the Revolution supplied a sense of an heroic past, while Dutch folklore pushed "history" back another century. And finally, the wild scenery of the Highlands provided a basis for national pride and self-confidence. Visitors from far and wide came to admire the Highlands. Their letters, books, sketches, and press commentary show that the Highlands' scenery and associations with the recent Revolution played a major role in defining American character and culture in the first half of the nineteenth century.

When the Erie Canal was completed, in 1825, it established the Hudson as a main artery of trade as well as travel, and launched a period of rapid economic expansion in New York. Evidence of increased activity was visible everywhere but nowhere more clearly than in the barges, paddlewheel steamers, and sloops which crowded the river, bedecked with banners and flags and engaging in frequent races. After the opening of the Erie Canal, traveling by waterways came very much into vogue. The Hudson was now a gateway to the West, making it possible to visit sights such as Niagara Falls.

Especially popular with Europeans and Americans alike was the "fashionable Northern Tour." The Tour typically departed from New York for America's most famous scenic spots, and West Point was the first stop. The journey from Manhattan took travelers past the basalt columns of the Palisades and the rolling meadows of Westchester until, three or four hours later, they reached the Highlands. There, they would visit the Point's historic sites, monuments, and famous scenic vistas. The tour continued on from West Point to Albany, often stopping for a side trip by stage coach to the Catskill Mountains. From Albany, travelers could go west on the Erie Canal to Niagara Falls, with excursions along the way to Oswego and the famous waterfalls at Genesee, near Rochester. Less rugged tourists might travel north from Albany to Lake George and Saratoga, or they might head east by overland route to Boston, taking the stagecoach from Albany for tours of Mount Holyoke and the Connecticut Valley or to the White Mountains and Lake Winnepesaukee.[7]

James Silk Buckingham, an English editor and member of Parliament, was one of many European travelers who took the American tour. His book, *America, Historical, Statistical, and Descriptive*, which was published in London in 1841, offers the following description of the journey from New York to West Point. It demonstrates the Europeans' overwhelming curiosity about the American landscape and its history. It also shows the sympathy and admiration felt even by the British for American independence.

> On the morning of Saturday, the 23rd of June (1838) we accordingly embarked at seven o'clock, on board the steamer for Albany, and found there between four or five hundred passengers bound up the river. . . . Leaving the wharf at the foot of Barclay Steet, we proceeded upwards on our course. . . . We passed the hills of Hoboken on our left; scattered over which, were many beautiful villas, the country-seats of opulent merchants. . . . About two miles beyond this, and eight from New York, the western bank of the river begins to assume a very remarkable appearance, presenting all along, on that margin of the stream, a perpendicular wall of rock. . . . These cliffs extend for nearly twenty miles along the western bank of the Hudson, and are called "The Palisadoes". . . . The opposite or eastern bank of the river was only of moderate height, wooded, and dotted over with dwellings at intervals, so as to contrast with the western cliffs. . . . At the termination of the Palisadoes, the river, which hitherto continues its breadth of about a mile, suddenly expands to a width varying from two to five miles, and is here called Tappan Bay. . . . About twenty miles beyond the bay of Tappan, and forty from New York, the scenery of the river becomes changed again, and the range of hills, called the Highlands, approach close to the water, and hem in the stream on either side. . . .

> The hills . . . present, for a distance of nearly twenty miles, a succession of bluff headlands or promontories, . . . and the ravines or valleys between them are as beautiful as the hills themselves. The windings round the promontories present a series of lakes, in which the spectator seems land-locked, as the continuation of the river is not visible either above or below, from the overlapping or interlacing of the headlands with the projecting capes of the other. This is particularly the case at a spot called the "Horse Race". . . . Here, too, the recollections of the revolutionary war are preserved in the names of Fort Montgomery and Fort Clinton, which were captured from General (Israel) Putnam by the British troops in 1777; and in the name of a sheet of water in the rear of Fort Clinton, called "Bloody Pond," from the crimson tinge given to the waters by the number of slain thrown into it after the sanguinary battle and dreadful carnage, of which that fort was the scene.
>
> About half-past ten we arrived opposite to West Point, having performed the distance of fifty miles in about three hours and a half. . . . The position of the fortress and the batteries of West Point, gave them a complete command of the river up and down, as far as the range of the cannon could extend; and every effort of the British, during eight years of warfare, to wrest them from the brave hands that defended both, were unsuccessful.[8]

Steamboat passengers approached West Point with hearts astir. Primed by stories and tales, travelers allowed their imaginations to roam through the mountain wilderness which rose up around them from the narrow, winding channel of the river, as they reached the southern gate to the Highlands. They scanned the peaks for signs of the supernatural creatures known to lurk there and listened for the rumbling spirits of the Dunderberg.

Almost certainly, as their steamboat neared Bear Mountain, the events of the Revolutionary War came to mind and perhaps scenes from *The Spy*, a current best-selling novel recalling the heroic but ill-fated defense of the Popolopen forts and the presence of George Washington and his troops.

In later years, when guidebooks were common, travelers read stories of the Revolution designed to stir their pride and admiration as they passed through the Highlands scenery. Such books told the lore of each place. Nearing Bear Mountain and the Popolopen gorge, they were reminded of the fortuitous escape of General Clinton and how, after the capture of the forts, the Americans burned their frigates to prevent those, too, from falling into enemy hands. Passing Beverly Dock, the scene of Benedict Arnold's flight to the safe haven of the British ship *Vulture*, they read of George Washington's despairing words: "Whom then, can I trust?" Every feature of the landscape took on new meaning, and many assumed almost mythical importance.

Frances Wright D'Auermont, an Englishwoman, expresses well the sentiment of many Europeans and Americans who visited West Point in this excerpt of a letter dated July 1819:

> It is impossible to enter, for the first time, the romantic pass of the Highlands, and to rest the eye upon the interesting academy of West Point; perched upon the highest and most rugged pinnacles, without recalling the traditionary and

Currier and Ives, *Cozzen's Dock, West Point,* c.1851 engraving showing the new Cozzen's hotel. Courtesy of the New-York Historical Society, New York City.

historical remembrances of the place. In earlier ages this was the region of superstitious terror to the Indian, and even the European hunter. The groans of imaginary spirits changed in time into the shrill pipe of war, and now it is only the mimic drum of the academy that rings among the caverns and precipices, through which the Hudson rolls his deep and confined waters.

It was in the fastness of West Point that, in the moment of his country's worst distress, the traitor Arnold planned his scheme of treachery.[9]

Steamboat passengers disembarked at Cozzens Landing in Highland Falls or were carried a short distance further to the West Point dock just below the plain, where they boarded a carriage which drove past the cadet encampment to the West Point hotel. West Point did not assume its present imposing military appearance until the turn of the twentieth century. In the early 1800s it was much smaller in scale and park-like in its arrangement of monuments, parade grounds, officers quarters, and barracks. Arriving there, guests freshened up and then set off for a walking tour of its famous scenery and historic monuments.

William Bartlett, *"Kosciusko's Monument,"* 1840 engraving from *American Scenery* by N.P. Willis. Originally published by George Virtue, London and R. Martin & Co., N.Y. 1840.

Kosciusko's Garden was a popular first stop. This hidden terrace overlooking the Hudson was constructed by the Polish general who came to this country to fight for the American independence and ultimately planned the fortifications at West Point. His memory was cherished by the American people, and nearly every nineteenth-century guidebook features this garden, which was his sanctuary for meditation. Visitors would seek out this secluded spot to recapture the mood of contemplation the general was known to have sought as wartime events swirled around him.

"The morning afterward cleared, and I went to Kosciusko's Garden," wrote Harriet Martineau, a controversial British social critic who visited West Point in 1835.

> I loved this retreat at an hour when I was likely to have it to myself. It is a nook scooped, as it were, out of the rocky bank of the river, and reached by descending several flights of steps from the platform behind the hotel and academy. Besides the piled rocks and the vegetation with which they are

Currier and Ives, *The Hudson from West Point*, 1862 colored lithograph showing one of the world famous views enjoyed by tourists at West Point. To the north, beyond the steamboat is the "northern gate" to the Highlands with Storm King on the west shore and Breakneck on the East. Newburgh Bay is visible in the distance. Courtesy of the New-York Historical Society, New York City.

> clothed, there is nothing but a clear spring, which wells up in a stone basin inscribed with the hero's name. This was his favorite retreat; and here he sat for many hours in a day with his book and his thoughts.

Dade's monument, Kosciusko's monument, and the Cadet Chapel were other favorite scenes for visitors. In the Chapel, they searched among the tablets on the wall commemorating Washington's generals for the "blank"—a stone with rank and date of birth but bearing no name—the tablet for Benedict Arnold, whose name was erased for his treachery. From the chapel, tourists continued to the adjacent cemetary. Located on bluffs above the Plain, it offered an unparalled opportunity to contemplate the valor of the nation's expired heroes amid awe-inspiring scenery. "Never did I see such a spot to be buried in," remarked Martineau.[10]

World-famous views could also be seen from the Siege Battery which overlooked the northern gate to the Highlands—Storm King and Breakneck—

and the broad expanse of Newburgh Bay beyond. James Silk Buckingham described them as "Two frowning Hills (which) overhung the stream on either side," a scene "of great grandeur and beauty," beyond which "the character of the landscape changes into a softer or more subdued style."[11] The evening parade and drills at sunset provided a fitting conclusion to a visitor's day.

As steamboating became ever more popular, a new genre of guidebooks developed to enhance the experience of travelers. To capture the interest of an increasingly affluent and mobile population, these books combined history, aesthetic commentary and even social gossip. Popular works included William Colyer's *Sketches of the North River* (1838) Benson Lossing's *The Hudson from Wilderness to the Sea* (1866) and Wallace Bruce's *The Hudson by Daylight* (1873).

Many of the travel books actively promoted the Highlands as a mecca for tourists, often providing basic information on where to go, where to stay, what to look for and how much it would cost. Wallace Bruce's *The Hudson by Daylight* advises the reader that round-trip excursion tickets from New York to West Point are "only $1.00, via Day Line Steamers." Fort Putnam, he adds "is a ruin worthy of a visit."[12] Such books often contained passages comparing the Hudson Valley to the Rhine and to famous European landscapes, as in this passage from William Colyer's *Sketches of the North River* (1838).

> *The Highlands*, the magnificent Matewan mountains of the Indians are now before us—they rise to the height of from twelve to fifteen hundred feet in bold and rocky precipices that would seem to defy the labor of man to surmount, and where the eagle builds her eyrie, and the hawk raises her callow brood, fearless of the strategems of the spoiler. The mighty river, pent within a narrow channel, struggles around the base of the hills, and lying deep in the shadows of the mountains, often appears like some dark lake shut out from the world. Our country offers no scenes more grand and sublime, and we may well doubt where the far-famed Rhine the cherished theme of every European tourist, can exceed in those attributes the one we are now voyaging upon. There is here, it is true, no classic spot to recall the days of the Caesars, "no castled crag of Drachenfels" encircled with its thousand years of romance, as on the German river, but to an American it is a land of enduring interest; every spot of this mountain pass, whereon man could gain a foothold, is interwoven with the history of a struggle that gave birth to a mighty empire, and, in coming ages, may we not suppose that it will be looked upon with some such feeling of veneration as with which we of the present now contemplate the pass of Thermopylae or the plains of Marathon?[13]

This type of comparison between America and Europe was not uncommon. Nationalist sentiment was so strong in favor of American scenery that books and articles frequently contained passages proclaiming this country's superiority. "Nature has wrought with a bolder hand in America," wrote N.P. Willis in his book *American Scenery*. Philip Freneau proclaimed that the Mississippi was a "prince of rivers in comparison to whom the *Nile* is but a small rivulet, and the *Danube* a ditch." "A fig for your Italian scenery!" Lanman exclaimed, ". . . this is the country where nature reigns supreme." The more moderate William Cullen Bryant suggested that "the southern shore of the Hudson" was "as worthy of a pilgrimage across the Atlantic as the Alps themselves."[14]

Even so, the Highlands may never have become the focus of major attention without the assistance of Sylvanus Thayer, Superintendent of West Point from 1817 to 1833. Sylvanus Thayer's name has been immortalized in Thayer Hall, the Hotel Thayer, the Thayer Gate and the Thayer monument. He is known as the Father of West Point—the hard-nosed disciplinarian who instituted the rigorous program of study for which West Point was to become famous, and the strict controls on cadet activity which, with some modification, continue to this day. Under his guidance, West Point emerged as a leader in civil engineering, training the men who built America's canals, railroads, and lighthouses. Less known is Thayer's role in making West Point one of the most important social and cultural centers in the country—a gathering place for artists, writers, politicians, merchants, industrialists, and distinguished guests from abroad—and a focus of nature-appreciation in America. Most likely, his only goal was to put West Point on firm footing as a military school, but the effect of his program was to draw international attention to the Highlands and their scenery.

Thayer was an astute politician and understood the relationship of public opinion to appropriations from Congress. Although he benefited from the strong support of President Madison, he knew better than to rely on that alone to aid him in building a strong military institution. He had many ways of cultivating support. One of them was to capitalize on the growing interest in the scenery and history of the Highlands.

In expanding the facilities at the academy, one of the most important things Thayer did was to build a hotel. The affect on the Highlands of this simple act was profound. From mere scenery to ride through, the area changed into a popular destination with a place to stay. The hotel had little to do with training for warfare but a great deal to do with developing a constituency, as it allowed day-boat passengers to spend more time at the academy and become acquainted with its many monuments, scenery, and, incidentally, its academic program. Thayer also built a residence for himself where he entertained distinguished guests, and he used his office as Superintendent to bring influential members of society to West Point.

Developing the academy as a first-class school of engineering produced public relations benefits as well, because it attracted an elite corps of cadets and drew the affection and interest of their families. Thayer hired a distinguished faculty and revived the Board of Visitors, an independent review team established to report to the War Department on conditions at West Point. Appointing to the board such members as Governor DeWitt Clinton of New York and Governor Oliver Wolcott of Connecticut, he assured that powerful men would be frequent visitors. The presence of influential people among the cadets, the faculty, and the board lured prominent guests and visitors. Authors Mark Twain, James Fenimore Cooper, and Charles Dickens, opera singer Jenny Lind, painter Frederic Church, dancer Fanny Elssler, Emperor Dom Pedro of Brazil,

Robert W. Weir, *Sylvanus Thayer*, 1843 oil on canvas. Weir was professor of drawing at West Point during Thayer's term as Superintendent. Courtesy of West Point Museum.

as well as Lord Napier and the Grand Duke of Russia are among the many names listed on the hotel guest registers at West Point.

Thayer succeeded not only in building support for his academy but also in promoting the historic and scenic values of the grounds. Through the diaries of visitors, the columns of journalists, and the travel accounts of authors and writers, the beauty and history of the Highlands reached a public who had never ventured there in person.

The Superintendent, the officers, and professors saw to it that guests were well entertained. In her 1842 article for readers of *Graham's Magazine*, Eliza Leslie describes the activities arranged for the pleasure of visitors. Music, dancing, sumptuous dining, and daily enjoyment of the romantic scenery of the Highlands were standard fare, punctuated by events such as concerts by distinguished vocalists, Thursday-night reading parties, and weekly chess games.

The social season opened on July 4 and continued through August while the cadets were in their summer encampment. The sights and sounds of cadet drills enhanced the pleasure of visitors, who never failed to turn out for the evening parade. The annual cadet ball was the gala event of the season, but for most of the summer, guests spent their days and evenings enjoying the scenery.

Miss Leslie informs us that the band was frequently enlisted for special moonlight serenades. A favorite performance on summer evenings was the *Nightingale*, a composition played under stately elms from whose branches a flute soloist imitated the warbling of birds. She describes another occasion when the "gentlemen attached to the military academy had made arrangements for taking the ladies on a moonlight voyage through the Highlands." Seven boats were assembled to carry the professors, officers, and the ladies, with soldiers at the oars. An eighth boat was appropriated to the band. "In the course of our little voyage," she writes, "several steamboats passed us: and all of them slackened their steam a while, for the purpose of remaining longer in our vicinity that the passengers might enjoy the music. One of these boats, in stopping to hear us, lay directly on the broad line of moonlight that was dancing and glittering on the water, the red glare of her lanterns strangely mingling with the golden radiance beneath. Our band was just then playing the Hunter's Chorus." The group rowed north as far as Storm King and then "the men resting on their oars, we floated down with the tide nearly as far as the Dunderberg, and never did this picturesque and romantic region look more lovely."[15]

The most prominent guests were entertained at the Superintendent's Quarters, and on Saturday evenings they rowed across the river to the weekly open house of the rich and hospitable industrialist Gouverneur Kemble. It has been said of the Superintendent's Quarters that it has welcomed more distinguished visitors than any other home in the United States except the White House.

The cadets were another attraction for visitors. Many of them were deemed to be very eligible bachelors, and West Point was a very popular place for upper-class tourists to bring their daughters. Political patronage dominated the appointments to West Point, with one cadet selected from each congressional district on the recommendation of the district's representative. While some candidates, generally westerners, were chosen on the basis of poverty or merit, many were the sons of eastern and southern aristocrats. For leading families, the academy was the place to send their second son, the eldest being trained to run the family business.

For nearly a century, West Point was the target of newspaper stories which charged the Army with perpetuating American aristocracy to the detriment of democratic values and the development of a competent military corps. The academy vigorously denied the characterizations of the press, but the charges contained more than a grain of truth. Regardless of birth, however, all soldiers pursued the same rigorous program. Cadet William Dutton complained in a letter to his cousin in 1843: "All I want of those Editors who say—'that lily fingered cadets, lounge on their velvet lawns—attend their brilliant balls & take pay for it' as I saw in a paper yesterday—is that they may go through but one 'plebe' encampment."[16] The result of this military training was, in the words of one observer, that the academy made "gentlemen of many intelligent youths, sprung from the humble grades of our people. It has made *men* of many scions of high estate whose talents would otherwise have been smothered under the follies of fashion and the enervations of luxury."[17]

The routine of the cadet contrasted starkly with the leisurely life of the vacationers at West Point. Regardless of background, certain rules applied to all soldiers at the academy. They were allowed no liquor, no tobacco, and no money. In four years at the post, they could not leave except for a two-month furlough at the end of their second year. During the summer encampment, there were drills from reveille at 5:00 A.M. until taps at 9:30 P.M., broken only by morning and evening parades, brief meals, and dancing class (without ladies) for an hour in the afternoon. The rest of the year was similarly regimented, except that scientific and engineering studies were substituted for drills, and study hall replaced dancing classes in the afternoon. Cadets described the experience in terms of slavery. They were not allowed to play cards, read novels, swim in the river, or even play a musical instrument.[18] The punishment for such infractions was dismissal from the academy or extra tours of duty.

Less than half the cadets appointed to the academy graduated. Among the casualties were Edgar Allan Poe, who walked off one day in 1831 without explanation and never returned, and James McNeill Whistler, noted later for a painting "Arrangement in Black and Grey: the Artist's Mother," popularly known as "Whistler's Mother." Whistler was dismissed in 1853 for an incorrect answer in chemistry class.

Whatever the miseries of cadet life for cadets, Thayer's rigorous program served only to enhance the pleasure of visitors. On tours of the grounds, guests enjoyed the sights and sounds of military training. The Irish airs and drill music of the band, the explosion of cannon, the daily maneuvers of brightly uniformed soldiers against a backdrop of white tent rows, and the sunset parades—all provided excellent atmosphere for viewing historic monuments and spectacular scenery.

Being part of the stage set did not prevent the cadets from enjoying the Highlands themselves. Letters they wrote described rambles through the woods which offered welcome respite from the regimented life on the post. On August 21, 1829, cadet Jacob Bailey wrote in a letter to his brother: "There is scarcely a place near the Point which I have not visited, from the highest point of the Crows Nest, to the muddiest marsh on the shores of the Hudson."[19] About a week later, he wrote again to his brother: "I went up to Crow's nest last Saturday and found a party of 10 or a dozen Cadets already on the top. When they started to come down, I appointed myself pilot and for my own amusement led them home by one of the most frightful ways which I knew, there is in reality not much danger in the path I chose, but it would make one not used to climbing feel somewhat queer. I pretended to lose my way and led them to the brink of a precipice some hundred feet high."

In 1823, Cadet Abner Hetzel wrote to his brother: "The Point & Country adjoining it far exceeds the most Sanguine expectations I had formed of it. I had anticipated entering a wilderness where there was nothing to gratify the optical sense."

Many of them carried images of the Highlands with them long after they left. Scholars have speculated that the following passage from Poe's tale, "Domain of Arnheim," describes the scenery which he enjoyed in his brief career as a West Point cadet.

> The usual approach to Arnheim was by the river. The visitor left the city in the early morning. During the afternoon he passed between shores of a tranquil and domestic beauty, on which grazed innumerable sheep, their white fleeces spotting the vivid green of rolling meadows. By degrees the idea of cultivation subsided into that of merely pastoral care. This slowly became merged in a sense of retirement—this again a consciousness of solitude. As the evening approached, the channel grew more narrow; the banks more and more precipitous; and these latter were clothed in richer, more profuse, and more sombre foliage. The water increased in transparency. The stream took a thousand turns, so that at no moment could its gleaming surface be seen for a greater distance than a furlong. At every instant the vessel seemed imprisoned within an enchanted circle, having insuperable and inpenetrable walls of foliage, a roof of ultra-marine satin, and *no* floor. . . . The channel now became a *gorge*. . . . The crystal water welled up against the clean granite, or the unblemished moss, with a sharpness of outline that delighted while it bewildered the eye.[20]

While Superintendent Thayer is credited with developing the code of cadet life and the program of study which made West Point famous, it was Superintendent Major Richard Delafield who did the most to change the post's physical appearance. In 1838, when Delafield assumed control, the academy was expanding, and Congress provided a substantial budget for constructing new buildings. Delafield selected the English Tudor style of architecture, establishing the military motif, and designed several of the structures himself.

The choice of a gothic style for West Point enhanced the region's romantic image. English Tudor developed out of its military function in medieval England, when warring barons built fort-like homes. Battlements, corner towers, moats, and narrow windows are some of the features which contributed to a structure's defensive strength. They were later refined into more decorative characteristics which flourished during the reign of the Tudor monarchs. In the first building Delafield designed, the Library, the military character of West Point is emphasized in the use of castellated towers and battlements. The

L. Behrens and Fritz Meyer, *View of West Point, United States Military Academy,* 1857 engraving. Courtesy of the Special Collection Division, U.S. Military Academy Library, West Point.

building blends with its surroundings through the use of native stone. The "First Division" of the Central Barracks, featuring battlements, drip stones and Gothic moldings, and the Dean's Quarters, a cottage with high gabled roofs, dormers and barge boards, are two remaining examples of the Delafield Tudor era.[21] If anything, the architecture of the Delafield era led to further comparisons between the Hudson Highlands and the Rhine Valley's "castled crags" and added to the pleasure of visitors.

However, no imitation of medieval European fortresses could compete with West Point's authentic ruin, Fort Putnam. That citadel, perched on the highest point of the academy grounds and offering the most spectacular views, was the favorite of visitors. Although no battles were ever fought there, it became a symbol of the American Revolution and a classic outpost for viewing the Highlands gorge. So it was natural that actress Fanny Kemble, recently arrived from Britain, would be escorted there on first arriving at West Point. The scene moved her to tears. It was one, as she noted in her diary, which should inspire the palette and the pen:

> Where are the poets of this land! Why such a world should bring forth men with minds and souls larger and stronger than any that ever dwelt in mortal flesh. . . . Have these glorious scenes poured no inspiration into hearts worthy to behold and praise their beauty? Is there none to come here and worship among these hills and waters till his heart burns within him, and the hymn of inspiration flows from his lips, and rises to the sky? . . . Yet 'tis strange how marvellously unpoetical these people are! How swallowed up in life and its daily realities, wants and cares; how full of toil and thrift, any money-getting labour. Even the heathen Dutch, among us the very antipodes of all poetry, have found names such as the Donder Berg for the hills, whilst the Americans christen them Butter Hill, the Crow's Nest, and *such like*. Perhaps some hundred years hence, . . . America will have poets."[22]

In fact, by 1832, the time of Kemble's visit, poets and painters were beginning to be recognized, and they took as their subject matter the very ground on which Kemble stood. Nationalism and interest in scenery had combined to produce new art forms for which America would soon be famous.

4. Landscape and Dreamscape: Hudson River Painters and Knickerbocker Writers

In the autumn of 1825, while Sylvanus Thayer was still in charge at West Point, a young man named Thomas Cole visited the military post. He was at the time neither powerful nor famous, and it is unlikely that he was entertained at the Superintendent's Quarters, or that his visit was even noticed.

Cole, an itinerant artist, had scraped together a living painting portraits and designing wallpaper to support his real passion—landscape painting. He had recently moved to New York City, and the scenery of the Hudson Valley captured his imagination. As a friend later wrote, "from the moment when his eye first caught the rural beauties clustering around the cliffs of Weehawken, and glanced up the distance of the Palisades, Cole's heart had been wandering in the Highlands, and nestling in the bosom of the Catskills."[1] At the first opportunity, Cole embarked on a sketching trip up the Hudson Valley. He was then twenty-four, and he visited the places which so intrigued him—the Palisades, the Highlands and the Catskills.

On reaching West Point, Cole rambled through the ravines and forested hills above the post. In all likelihood, he carried with him his writing and sketching pads and his flute, pausing here and there to play a few notes as he sought to capture the mood of the place he intended to paint. Climbing to the highest vantage point, he probably pulled out pen and paper, as was his habit, to record his observations:

Asher B. Durand, *Kindred Spirits,* 1849 oil on canvas, showing nature poet William Cullen Bryant and Hudson River School artist Thomas Cole in a setting in the Catskill Mountains, a work commissioned by art patron Jonathan Sturges upon the death of Thomas Cole. Courtesy of the New York Public Library.

"Mists were resting on the vale of the Hudson like drifted snow," he once wrote of a river scene. "Tops of distant mountains in the east were visible—things of another world. The sun rose from bars of pearly hue. . . . The mist below the mountain began first to be lighted up, and the trees on the tops of the lower hill cast their shadows over the misty surface—innumerable streaks." He would refer to such notes later, when he returned to his studio to paint.

Of particular interest to Cole on this hillside were the ruins of Fort Putnam. Crumbling and overgrown with vines after forty years of neglect, the fort brooded over a panorama of spectacular beauty—bold and rugged mountains, the dark river far below and in the distance, the peaks of the Catskills, charcoal blue under autumn's gathering clouds. From this granite promontory, rich with historic associations, Cole, like many others who visited West Point in the early nineteenth century, must have felt close to God and country.

From West Point, Cole continued his journey north to the Catskills, where he traveled to Kaaterskill Falls and other nearby wilderness spots. Later, when he returned to his cramped apartment on Greenwich Street in New York, he painted the scenes which had impressed him on his trip—each a dramatic portrait of untamed wilderness. A framemaker put three of the canvasses on display in his shop window: one, a view of an unnamed locality, titled *Lake with Dead Trees*; another, a Catskill Mountain scene; and the third *A View of Fort Putnam* in the Hudson Highlands.

Passing by the framemaker's shop one day was one of New York's leading artists, John Trumbull, who spied Cole's work and admired it. Trumbull purchased one of the paintings, *The Falls of the Caterskill*, which he hung in his own studio. The same day, he invited his friends, fellow artists William Dunlap and Asher Durand, to come see Cole's paintings and make the acquaintance of the artist, whom Trumbull had invited to his studio.

The encounter was embarrassing for Cole, who was described by his biographer Louis L. Noble as a slight young man whose eyes shone with a combination of eloquent brightness and feminine mildness. Cole was tongue-tied and nervous. He didn't know what to say when Trumbull exclaimed, "You surprise one, at your age, to paint like this. You have done what I, with all my years and experience, am yet unable to do."

Despite his reticence, Cole was pleased with the results of the encounter. Durand immediately purchased the *View of Fort Putnam*, and Dunlap purchased the third canvas. They paid $25.00 each. Dunlap soon sold his to Mayor Philip Hone, one of the city's most important collectors and a patron of the arts, who offered him $50.00 for the painting. Dunlap later wrote: "My necessities prevented me from giving the profit, as I ought to have done, to the painter. One thing I did, which was my duty. I published in the journals of the day an account of the young artist and his pictures; it was no puff but an honest declaration of my opinion, and I believe it served merit by attracting attention to it." With such important backing, Cole's life and fortunes were transformed almost overnight. "His fame spread like fire," said Durand, and an exuberant American cultural movement was born.

Cole's solitary venture into the Hudson Valley wilderness and the discovery of his work launched a celebration of nature in art that became known as the Hudson River School of Painting. Following Cole's lead, scores of artists would make pilgrimages in search of spectacular scenery to paint. Like Cole, they found inspiration in the Highlands' crumbling ruins, rugged mountains, and dark river.

What Thomas Cole did was present nature in its raw beauty, full of the unexpected, the dramatic, and the intimate. His work was a departure from the accepted technique of his time, and, unlike the literal pictures of his contemporaries, it captured a mood and a sense of nature's mystery. Not meant to portray a specific town or place, his canvases seemed to speak of Divine Creation. Twisted, wind-swept trees evoked nature's untamed power. Streaks of ethereal light penetrated luminous clouds to awaken a landscape slumbering in shadow, preserving the ancient secrets of the primeval forest. In the Highlands, where civilization had made few inroads, Cole found such a landscape to paint.

This was a new concept in artistic expression. Most painting at the time portrayed the families of wealthy patrons (who commissioned the portraits as proof of their status and distinction) or depicted historical events, such as famous battles. Landscapes appeared primarily as engraved illustrations in travel books or on military maps.

Cole's work struck a responsive chord in his viewers, and he was soon emulated by Durand. Others who joined Cole on the roster of Hudson River School artists were John Casilear, John Frederick Kensett, Sanford Gifford, Thomas Doughty, George Inness, David Johnson, Thomas Rossiter, Jasper Cropsey, Robert Weir (Professor of drawing at West Point), and Frederic E. Church (a student of Cole). The term *Hudson River School* was first coined in a contemptuous article in the *New York Herald* meant to ridicule these landscape paintings.[2] However, the work of these and dozens of other less-known artists attracted international attention and acclaim for the next fifty years.

Their work was united not so much by a common style as by a shared view of nature and humanity's relation to it. They celebrated nature's grandeur in both its beautiful and its savage aspects, especially the vast, pristine wilderness. These artists associated unspoiled nature with virtue, and if their paintings show people at all, they are small, isolated figures dwarfed by dramatic, wild scenery. The scant signs of civilization show a time gone by. Pastoral scenes are common, with the symbolic cow grazing even on the rockiest mountain precipice. There may be the white sails of sloops, but few if any steamboats.

If the Hudson River School painters broke with artistic tradition, they were in perfect synchrony with the prevailing philosophy of Romanticism. By 1825, the ideas expressed in the writings of Rousseau in France and Sir Walter Scott in Britain, were beginning to take hold in America. In addition to glorifying an idealized past, these European writers found in nature a purity combined with a mysterious disharmony which spoke to them of a higher being. In their view, nature was God's finest work. The power of nature to inspire terror and awe transported the nineteenth-century Romantics to a plane of religious experience. In the Highlands, they found a landscape where God's work in nature was manifest. The mountains, the river, the wind, and the rocks moved them as would a sermon.

Fanny Kemble, the British actress, visiting West Point in 1832, expressed such feelings in her diary as she recounted her experience hiking up to the ruins at Fort Putnam—the same place painted by Cole:

> Saturday, November 10, 1832 . . .
>
> Alone, alone, I was alone and happy, and went on my way rejoicing, climbing and climbing still, till the green mound of thick turf, and ruined rampart of the fort arrested my progress. I coasted the broken wall, and lighting down on a broad, smooth table of granite fringed with young cedar bushes, I looked down, and for a moment my breath seemed to stop, the pulsation of my heart to cease—I was filled with awe. The beauty and wild sublimity of what I beheld seemed almost to crush my faculties—I felt dizzy as though my senses were drowning—I felt as though I had been carried into the immediate presence of God. Though I were to live a thousand years, I never can forget it. The first thing that I distinctly saw, was the shadow of a huge mountain, frowning over the height where I stood. The shadow moved down its steep sunny side, threw a deep blackness over the sparkling river, and then passed off and climbed the opposite mountain on the other shore, leaving the world in the full blaze of noon. I could have stretched out my arms and shouted aloud—I could have fallen on my knees and worshipped—I could have committed any extravagance that ecstacy could suggest. I stood filled with amazement and delight, till the footsteps and voices of my companions roused me. I darted away, unwilling to be interrupted.[3]

Two of the most important concepts in the philosophy of Romanticism were the "sublime" and the "picturesque." Few words were used more frequently to describe the Hudson Valley, particularly the Highlands and the neighboring Catskills.

In reaction to the classical concept of beauty, which was logical and harmonious, the *sublime* and the *picturesque* were terms adopted by the Romantics to describe the irregular and unpredictable aspects of nature. The term *sublime* was reserved for the more violent aspects of nature: thunder storms, rushing waterfalls, dense and tangled vegetation, plunging cliffs and cosmic sky.[4] The *picturesque* differed only in degree, being tamer and more sentimentalized. Irregularity of form, texture, or detail characterized the picturesque.

For nineteenth-century Romantics, viewing sublime scenery was a spiritual act. Thomas Cole commented on this in his 1835 "Essay on American Scenery": "Those scenes of solitude from which the hand of nature has never been lifted," he wrote, "affect the mind with a more deep toned emotion than aught which the hand of man has touched. Amid them the consequent associations are of God the creator—they are his undefiled works, and the mind is cast into the contemplation of eternal things."

For the artists of the Hudson River School, a painting was, in a sense, a portrait of God. A landscape painting "will be great in proportion as it declares the glory of God, by a representation of his works,"[5] declared Asher Durand.

The sublime and picturesque qualities which the painters sought to capture on canvas could be readily found in the Hudson River Valley scenery. The Highlands, particularly, conformed to the Romantic ideal of sublime landscape, with craggy precipices, the tortured course of the river as it entered the Northern gate, and the wind-swept trees growing from barren rock. The Catskills, on the river's western horizon, also fit this image, although the two regions were portrayed in different ways.

In the Highlands, where the mountain passes were enshrouded with associations reminiscent of the recent War for Independence, a historical element prevailed. This evocation of an idealized past was popular with the Romantics. The most favored scene was the Storm King Gorge, looking to the north viewed from Fort Putnam or looking south from Newburgh or New Windsor. Bull Hill (Mt. Taurus) and Anthony's Nose, as well as views of Cold Spring and the surrounding scenery, were other favorites (see color plates). Often these views included the picturesque West Point encampment as well as sloops, and occasionally steamboats. Images of the Catskills, on the other hand, tended to show a remote and ancient wilderness of rocky ledges, cascading waterfalls and dark, forested mountains.

To a receptive public, the work of the Hudson River School artists established the Highlands as hallowed ground and helped to popularize the area for tourism. Travelers like Fanny Kemble came to West Point not only as sightseers but also as worshippers at nature's shrine.

The artists of the Hudson River School sought scenes that looked like paintings—scenes that provided all the elements of Romantic vision. Where nature did not comply, the artists sometimes invented. For example, the painter might add a northward setting sun and celestial rays of light to illuminate the Highlands' mountain scenery and imply a heavenly presence. Nature's grandeur was suggested by the extension of mountains, clouds, and rivers beyond the observer's vison, with small human figures, backs turned, idling in the foreground, leading the viewer to identify with them and emphasizing the sensation of vastness and isolation. The Highlands gorge, close, confined, and dark, opens out into Newburgh Bay, where the Catskills can be seen far in the distance. In this scene, such an effect could easily be captured by the painter.

Painters traveled widely in search of awesome and sublime wilderness, and the Hudson was not their only subject matter. Their travels brought them to the Catskills, Lake George, the Connecticut Valley, the White Mountains of New Hampshire, Niagara Falls, and the Adirondacks. One of Frederic Church's most famous paintings is of the Andes in Peru. Yet in the Hudson Valley, where the long ribbon of river binds the Catskills and the Highlands, artists found some of the most dramatic scenery in eastern America. Cole always returned to it, his friend Noble said, "with the tenderness of a first-love."[6]

In the mid-nineteenth century, residing and painting in the Hudson Valley

Home of painter Thomas P. Rossiter in Cold Spring. Copyright © 1990 by Robert Beckhard.

was considered essential to art education in much the same way as living and working in Paris was the ideal for the next generation of artists. About five hundred oil paintings of the Hudson from this period have been preserved to this day, more than one hundred of them scenes of the Highlands, including works by almost all major artists.[7]

Many of the artists of the Hudson River School established themselves in estates and studios on the shores of the river they loved. The most elaborate is Frederic Church's moorish castle "Olana," itself a work of art, located high on a hill south of the city of Hudson, overlooking the river and the Catskills and now preserved as a state park. Thomas Rossiter built a palatial mansion in Cold Spring, Thomas Cole set up a studio in Catskill, Asher Durand painted from New Windsor, overlooking the Highlands, and Jasper Cropsey had a studio, "Ever Rest," in Hastings-on-Hudson. Albert Bierstadt had a great mansion on the river near Tarrytown, and Robert Weir painted from West Point.

The achievement and world acclaim of the Hudson River School would not have been possible without the support of a growing merchant class. The opening of the Erie Canal in 1825 brought great prosperity to New York City, and it became fashionable for merchants to display their wealth through patronage of American artists. These businessmen not only maintained impressive art collections but also commissioned major works and sent artists abroad to study. Jonathan Sturges, a New York merchant, financed Durand's 1840 trip to Europe where he joined Kensett, Casilear, and Rossiter in copying old masters and painting Swiss and Italian landscapes. Cole had several important backers, among them Mayor Philip Hone, Luman Reed, a wholesale grocer and business partner of Sturges, and Thomas Gilmor of Baltimore, who loaned Cole $300 in 1829 so he could travel abroad. The increase in the number of patrons was accompanied by a major change in taste. The new merchants desired greater artistic variety, and this allowed the Hudson River School to flourish.

The Hudson River Painters were aided by the Knickerbocker writers, so-called after Washington Irving's fictional *Knickerbocker History of New York,* who, a decade before Cole's historic sketching trip, had made the American landscape a fashionable topic of discussion. By 1825 these writers were in a position to promote the Hudson River School artists' work through articles in the popular press and later through mutual publications. Influenced by the same romantic concepts, the Knickerbockers were able to express the moral and spiritual significance of nature in a way that the visual artists could not. By the time the term Hudson River School was coined, the Knickerbockers had already published major works which popularized nature in America and paved the way for acceptance of the painters' break with artistic tradition.

The Knickerbockers began to be published in the same year that the steamboat *Clermont* was launched. With intense national and international interest in the Hudson and the Highlands resulting from steamboat travel, it is no coincidence that the region appeared frequently as a subject and a setting in this new literary genre. The work of the Knickerbockers created a body of legend, poetry, caricature and supernatural lore about American scenery and landscape which helped to enhance the Hudson Valley's romantic image. In this, the Highlands, steeped in folklore, familiar as a battleground of the Revolution, and dramatically beautiful with its wild, raw, scenery, played a major role.

Technically the term *Knickerbocker* could apply to any author working out of New York City. Washington Irving, James Fenimore Cooper, and William Cullen Bryant were the brightest lights. The "Knickerbocracy," however, included some twenty lesser luminaries, such as Nathaniel Parker Willis, Fitz-Greene Halleck, Joseph Rodman Drake, Washington Irving's brother William and James Kirke Paulding—writers who were popular in their day but whose reputations have faded with time. Melville, Poe, and Whitman were also New Yorkers, but their work was part of a later tradition. The reputation of America's first group of literary romanticists waxed in 1807 with the publication of *Salmagundi,* a collection of witty prose and verse by the two Irving brothers and Paulding. It began to wane by the 1850s, when the new works of Emerson, Melville, and Hawthorne rose to prominence.

During these four decades the Knickerbockers stamped a uniquely American imprint on a romantic style which made them as famous in the world of letters as the Hudson River School Painters were in the world of art. Their numbers and prestige, the literary magazines they produced (the *New York Mirror* and the *Knickerbocker*), and the publishing business they fostered wrested cultural pre-eminence from Boston and Philadelphia and established New York as the literary capital of America. Between 1820 and 1852, 345 publishers operated out of New York City, more than twice the number as in Philadelphia.

There was very little uniformity amongst the works of individual Knickerbockers. They used literary forms as diverse as epic poetry, satire, and the historic novel. What they shared, however, was a romantic vision and a nationalistic temperament.

Born on the eve of the nation's independence, tales of the Revolution and the legends of the Dutch fired their imaginations. Many of them took as their subject matter the history, landscape, and folklore of their native land. Through these works, they gave shape to the American character and helped define a national identity. The Highlands, which had so many of these elements, served them well as subject and setting, and in turn became symbolic of American heritage.

Washington Irving was one of the acknowledged leaders, and it was his first

Washington Irving at the age of 27.

significant book, *A History of New York by Diedrich Knickerbocker*, from which the group derived its name. Drawing on fragments of history and even greater amounts of fantasy and imagination, *Knickerbocker's History* used the Hudson Valley landscape as a backdrop for a tale which poked fun at the burghers who settled the new world. By making his readers laugh at the primitiveness of American life, he made it acceptable. Along with *The Sketch Book*, published a decade later, *Knickerbocker's History* was a popular and lasting favorite overseas, where curious European readers enjoyed the novelty of the American scene and the wit and satire Irving employed. *The Sketch Book* contains the famous stories "Rip Van Winkle" and "The Legend of Sleepy Hollow," and was so beloved that Lord Byron claimed to have memorized every word, and Sir Walter Scott to have laughed uproariously on reading it the first time.

The Highlands as well as the Catskills and Sleepy Hollow, a fictionalized version of an area near Tarrytown where Irving had a Hudson River country home called "Sunnyside," appeared repeatedly in these books as well as in *Bracebridge Hall*, one of Irving's lesser-known works. His work responded to the growing national interest in American scenery and sought, as Irving wrote in the introduction to *Knickerbocker's History*, to "clothe home scenes and places . . . with those imaginative and whimsical associations so seldom met in our new country, but which live like charms and spells about the cities of the old world, binding the heart of the native inhabitant to his home."

Each of the Knickerbocker writers presented the Highlands in a different way—some with great seriousness, some with humor and satire. They embellished the region's natural history, folklore, and Revolutionary War events to achieve the desired Old World effect. Irving's *Knickerbocker's History* was a burlesque of the local customs, institutions, and scenes of New York. In it, the author uses the Highlands to symbolize America's untamed wilderness, and in the process he endows the region with new "charms and spells." Written from the point of view of a New Yorker of Dutch descent, it covers New York's history "from the beginning of the world to the end of the Dutch dynasty." Several hilarious chapters cover the Governorship of Peter Stuyvesant, including an account of a voyage up the Hudson River with Antony Van Corlaer to muster troops for a war on the Swedes. As their galley approaches the Highlands, "where it would seem that gigantic Titans had erst waged their imperious war with heaven, piling up cliffs on cliffs, and hurling vast masses of rock in wild confusion," the author, Diedrich Knickerbocker, launches into a story intended to create new folklore about the Highlands with a satirical bent.

> And now I am going to tell a fact, which I doubt much my readers will hesitate to believe, but if they do they are welcome not to believe a word in this whole history—for nothing which it contains is more true. It must be known then that the nose of Anthony the trumpeter was of a very lusty size, strutting boldly from his countenance like a mountain of Golconda, being sumptuously bedecked with rubies and other precious stones—the true regalia of a king of good fellows, which jolly Bacchus grants to all who bouse it heartily at the

Title page of *Knickerbocker's History of New York* by Washington Irving.

A

HISTORY OF NEW YORK,

FROM THE

BEGINNING OF THE WORLD TO THE END OF THE DUTCH DYNASTY;

CONTAINING, AMONG MANY SURPRISING AND CURIOUS MATTERS, THE UNUTTERABLE PONDERINGS OF WALTER THE DOUBTER, THE DISASTROUS PROJECTS OF WILLIAM THE TESTY, AND THE CHIVALRIC ACHIEVEMENTS OF PETER THE HEADSTRONG—THE THREE DUTCH GOVERNORS OF NEW AMSTERDAM: BEING THE ONLY AUTHENTIC HISTORY OF THE TIMES THAT EVER HATH BEEN OR EVER WILL BE PUBLISHED.

BY

Diedrich Knickerbocker.

De waarheid die in duister lag,
Die komt met klaarheid aan den dag.

THE AUTHOR'S REVISED EDITION.

COMPLETE IN ONE VOLUME.

NEW YORK:
G. P. PUTNAM, 532 BROADWAY.
1861.

> flagon. Now thus it happened, that bright and early in the morning, the good Antony, having washed his burly visage, was leaning over the quarter railing of the galley, contemplating it in the glassy wave below. Just at this moment the illustrious sun, breaking in all his splendour from behind a high bluff of the Highlands, did dart one of his most potent beams full upon the refulgent nose of the sounder of brass—the reflection of which shot straightway down hissing hot into the water, and killed a mighty sturgeon that was sporting beside the vessel. This huge monster, being with infinite labour hoisted on board, furnished a luxurious repast to all the crew, being accounted of excellent flavour excepting about the wound, where it smacked a little of brimstone—and this, on my veracity, was the first time that ever sturgeon was eaten in these parts by Christian people. When this astonishing miracle became known to Peter Stuyvesant, and that he tasted of the unknown fish, he, as may well be supposed, marvelled exceedingly; and as a monument thereof, he gave the name of *Antony's Nose* to a stout promontory in the neighbourhood, and it has continued to be called Antony's Nose ever since that time.[8]

The story bears remote resemblance to the tale of the naming of Anthony's Nose found in the diary of Freeman Hunt and demonstrates Irving's personal familiarity with the Highlands and their folklore. He spent many weekends in Cold Spring at the home of his friend Gouverneur Kemble and, like many other Knickerbockers, he was a frequent visitor at West Point. His earliest exposure to the region was in 1800, when at age seventeen he traveled up the Hudson River by sloop. Years later, in his autobiography, he recalled the sounds of the leaping sturgeon and the song of the whippoorwill echoing in the night, as the sloop lay at anchor below those dark and mysterious mountains. It was at a time, he wrote, "before the steamboats and railroads had annihilated time and space," when a voyage to Albany "was equal to a voyage to Europe at present, and took almost as much time."[9]

By the time the steamboat became popular, other Knickerbockers were writing witty satire, and the traveling public was their audience. Anthony's Nose seems to have been a favorite subject because it was familiar through folklore and lent itself to literary embellishment for romantic appreciation of the American landscape. In his guidebook, the *New Mirror for Travellers*, Knickerbocker James Kirke Paulding recites a story on this topic, in a style which early nineteenth-century readers found highly entertaining:

> Turning the base of Donderbarrack (Dunderberg), the nose of all noses, Anthony's Nose, gradually displays itself to the enraptured eye, which must be steadily fixed on these our glowing pages. Such a nose is not seen every day. Not the famous hero of Slawkembergius, whose proboscis emulated the steeple of Strasburg, ever had such a nose to his face. Taliacotius himself never made such a nose in his life. It is worth while to go ten miles to hear it blow—you would mistake it for a trumpet. The most curious thing about it is, that it looks no more like a nose than my foot. But now we think of it, there is still something more curious connected with this nose. There is not a soul born within five miles of it, but has a nose of most jolly dimensions—not quite as

> large as the mountain, but pretty well. Nay, what is still more remarkable, more than one person has recovered his nose, by regularly blowing the place where it ought to be, with a white pocket handkerchief, three times a day, at the foot of the mountain, in honor of St. Anthony. In memory of these miraculous restorations, it is the custom for the passengers on steam boats, to salute it in passing with a universal blow of the nose: after which, they shake their handkerchiefs at it, and put them carefully in their pockets. No young lady ever climbs to the top of this stately nose, without affixing her white cambric handkerchief to a stick, placing it upright in the ground, and leaving it waving there, in hopes that all her posterity may be blessed with goodly noses.[10]

In a far more serious vein was the work of James Fenimore Cooper. His first bestseller, published in 1821, soon after Irving's *Sketchbook*, was *The Spy: A Tale of the Neutral Ground*, which takes place in the Hudson Valley during the Revolutionary War. Like Cooper's other works, the novel evoked images of heroic characters, telling the story of a double agent who nobly served the American cause. In this work, the Highlands evoke patriotic memories and remind the reader that this land is historic ground. Thought to be based on the life of an actual spy, Enoch Crosby, the tale follows the fictional spy, Harvey Birch, through British-held Westchester County and the neutral territory of the Van Cortlandt Manor to the American encampment in the Hudson Highlands. There, Birch meets up with George Washington and delivers him vital information. The book concludes with Washington offering Birch a hundred doubloons for his services. Washington attributes the small sum to "the poverty of our country" and cautions Birch that he must go to his grave with the reputation of an enemy to his native land, as to do otherwise would jeopardize the lives and fortunes of Washington's network of secret agents. Birch, however, refuses the reward, saying he exposed his life and "blasted his character" not for money but for love of country. Birch dies a pauper but confident of the special regard in which he was held by his country's founder.

The Spy glorifies the drama of the quest for independence and those who served the cause. Woven into the patriotic fabric of the story are repeated references to the bounty of the countryside and the beauty of the scenery. For Cooper, love of country was strongly linked to appreciation of its natural resources. The following is one of the book's many references to the Highlands and illustrates how Cooper uses the emotional effects of nature in telling the story of the quest for national independence. Cooper's descriptions were much like the paintings of his contemporaries and were intended to bring the reader closer to God and country.

> They soon reached the highest point in their toilsome progress to the summit, and Frances seated herself on a rock to rest and to admire. Immediately at her feet lay a deep dell, but little altered by cultivation, and dark with the gloom of a November sunset. Another hill rose opposite to the place where she sat, at no great distance, along whose stinted growth showed a meagre soil.
>
> To be seen in their perfection, the Highlands must be passed immediately after the fall of the leaf. The scene is then the finest, for neither the scanty

foliage which the summer lends the trees, nor the snows of winter are present to conceal the minutest objects from the eye. Chilling solitude is the characteristic of the scenery; nor is the mind at liberty, as in March, to look forward to a renewed vegetation that is soon to check, without improving the view.

The day had been cloudy and cool, and thin fleecy clouds hung around the horizon, often promising to disperse, but as frequently disappointing Frances in the hope of catching a parting beam from the setting sun. At length a solitary gleam struck on the base of the mountain on which she was gazing, and moved gracefully up its side, until reaching the summit, it stood for a minute, forming a crown of glory to the sombre pile.[11]

The work of Irving, Cooper, and other Knickerbockers cast a spell of history and legend which the public would not soon forget. No less important was their contribution to national thought about nature's spiritual value. Like the Hudson River School artists, the Knickerbockers were druids at heart. "The groves were God's first temples," wrote William Cullen Bryant in "A Forest Hymn."[12]

Bryant's poem "Thanatopsis," written at the age of seventeen, ushered in a new approach to the contemplation of nature in American literature. His sincerity and purity contrasted sharply with contemporary "poetical platitudes" and was a milestone in the history of American poetry. Published in 1817, the poem is a meditation on death and instructs the reader to seek mental and spiritual solace in nature. It begins:

To him who in the love of Nature holds
Communion with her visible forms, she speaks
A various language; for his gayer hours
She has a voice of gladness, and a smile
And eloquence of beauty, and she glides
Into his darker musings with a mild
And healing sympathy that steals away
Their sharpness ere he is aware. When thoughts
Of the last bitter hour come like a blight
Over thy spirit, and sad images
Of the stern agony, and shroud, and pall,
And breathless darkness, and the narrow house
Make thee to shudder and grow sick at heart,
Go forth, under the open sky, and list
To Nature's teachings, while from all around,—
Earth and her waters, and the depths of air,—
Comes still a voice . . .[13]

The Knickerbockers' work emerged at a time when a great intercontinental debate raged in the popular press, questioning the ability of America to produce anything of merit. Curiously, the widespread American and European acclaim enjoyed by this group of writers did nothing to stem the controversy. At first, the focus was on America's cultural and intellectual capabilities. Sydney Smith, writing in 1820 for the *Edinburgh Review* stated it bluntly: "In the four corners of the globe," he wrote, "who reads an American book? or goes to an American play? or looks at an American picture or statue?" This comment produced "paroxysms of wrath" in the American press, but to a certain extent this feeling of cultural inadequacy was shared by Americans and persisted for decades, long after the Knickerbockers and the Hudson River School painters had won wide acceptance in the United States and abroad.[14]

In 1817, when William Cullen Bryant submitted his poem "Thanatopsis" to the *North American Review*, the editor, Richard Dana, is said to have remarked, "No one on this side of the ocean is capable of writing such verse."[15] However, as comparisons between Europe and America continued, they extended beyond art and literature into the merits and defects of the American landscape. To Europeans who were accustomed to traveling in Greece, Egypt, Syria, and Germany's Rhineland, landscape was far more beautiful if cloaked in legend or improved by picturesque ruins or other signs of human endeavor. In this, America was considered seriously deficient, although some who entered into the debate, like Goethe and Chateaubriand, felt otherwise. In an 1827 poem addressed to the United States, Goethe expressed his opinion that traditions and historic associations were not necessary for the New World:

> America, you're better off than
> Our continent, the old.
> You have no castles which are fallen
> No basalt to behold.
> You're not disturbed within your innermost being
> By useless remembering
> And unrewarding strife.
> Use well the present and good luck to you
> And when your children begin writing poetry
> Let them guard well in all they do
> Against knight-robber-and ghost-story.[16]

Goethe's admonitions failed to silence the critics, and the debate continued for years to come. In 1835, painter Thomas Cole rose to the defense of "what has been considered a defect in American scenery—the want of associations such as arise in the scenes of the old world." The American scenes, he responded, "are not destitute of historical and legendary associations—the great struggle for freedom has sanctified many a spot, and many a mountain, stream, and rock, has its legend, worthy of a poet's pen or the painter's pencil."[17]

By the time Cole wrote his "Essay on American Scenery," the Knickerbockers had already ignored Goethe's advice and dignified many battlefields, mountains, streams, and rocks, especially in the Highlands, with volumes of legends, history, ghost-stories, and poetry which were widely read in Europe as well as America. Books by Irving, Cooper, and Bryant sold many copies abroad and contributed greatly to European interest in the new world.

Their work established the history that Europeans had found wanting in America. In the wild nature of the Highlands, the Catskills, and other parts of

New York and New England, they found a replacement for Old World ruins. "Those vast aboriginal trees," Sir Walter Scott commented to Washington Irving, "are the monuments and antiquities of your country."[18] The ruins of Fort Putnam, the folklore of Anthony's Nose and Dunderberg, the legends of the Revolutionary War, and the virgin forest of the Highlands offered the heritage which had seemed to be lacking.

Nevertheless the national inferiority complex lingered on. "The Culprit Fay" a poem about the Crow's Nest in the Highlands by Knickerbocker Joseph Rodman Drake is less notable for its verse than for the story which surrounded its creation. An 1894 steamboat guidebook by Wallace Bruce recounts that "It is said that the 'Culprit Fay' was written by Drake in three days, and grew out of a discussion which took place during a stroll through this part of the Highlands between Irving, Halleck, Cooper and himself, as to the filling of a new world with old-time legends."[19] According to the story, Drake called his friends together three days later and read them the poem which begins with these lines:

> Tis the middle watch of a summer's night,—
> The earth is dark, but the heavens are bright . . .
> The moon looks down on old Cro'Nest,
> She mellows the shades on his shaggy breast,
> and seems his huge grey form to throw
> In a silver cone on the wave below;[20]

The story was repeated in tourist guides along with these stanzas of the poem and a discourse on the similarity of American culture to that of Europe, Greece, and Rome. The use of the story illustrates the growing preoccupation with Knickerbocker literature as proof of the coming of age of American culture.

As if to emphasize a dependency on old world traditions, Drake, like other Knickerbockers, used European literary forms to celebrate his native land. "The Culprit Fay" is a lengthy epic tale in a romantic style. It tells the story of a fairy, or fay, who has broken his vows of chastity and fallen in love with an earthly maid. The poem takes place at midnight on the summit of Crow's Nest to the south of Storm King. The assembled spirits of the forest impose a penance whereby the "culprit fay" braves the cold waters of the Hudson, captures a drop of spray from the leaping sturgeon, and eventually succeeds in returning to the mountain top just as dawn is breaking, and so restores his fairy powers.

Knickerbocker essayist Nathaniel Parker Willis was equally concerned with creating a mantle of romantic associations in which to cloak American scenery. It was in a piece published in the *Home Journal* that he advocated changing the name of Butter Hill to Storm King.

> The tallest mountain, with its feet in the Hudson at the Highland Gap, is officially the Storm King—being looked to, by the whole country around, as the most sure foreteller of a storm. When the white cloud-beard descends upon his breast in the morning (as if with a nod forward of his majestic head), there is sure to be a rain-storm before night. Standing aloft among the other

> mountains of the chain, this sign is peculiar to him. He seems the monarch, and this seems his stately ordering of a change in the weather. Should not STORM-KING, then, be his proper title?[21]

The notion appealed to Willis's readers, and public sentiment led to the rapid adoption of the new name.

The Knickerbockers were all well acquainted with each other and were known to be, in general, a fun-loving group. They met frequently to consume great quantities of food and drink and amuse themselves in composing verse and bits of satire. A favorite haunt was the country estate of their friend Gouverneur Kemble, a businessman whose family had built a mansion on the banks of the Passaic near Newark. In the *Salmagundi* papers, Knickerbockers Irving and Paulding refer to themselves and their friends (including Henry Brevoort, Henry Ogden, John and Gouverneur Kemble, and Irving's two brothers William and Peter) as the "Nine Worthies" of "Cockloft Hall," their affectionate name for the Kemble estate. Of their days of fun and hilarity at the Hall, one of them recalled that "we sported on the lawn until fatigued, and sometimes fell sociably into a general nap in the drawing room in the dusk of evening." In town, they often assembled at Dyde's tavern on Park Row, or at a porter-house at the corner of Nassau and John Streets. One friend reported that after a festive evening Irving fell through an open grate on the way home. His solitude and depression was relieved after several guests joined him there, where they sat and laughed until dawn. Years later, Irving wrote to Kemble "Who would have thought that we should ever have lived to be such respectable old gentlemen?"[22]

The Knickerbocker writers also developed close alliances with Hudson River School Painters with whom they shared a romantic philosophy and a common interest in American landscape. The relationship between William Cullen Bryant, Thomas Cole, and Asher Durand was most significant. When Thomas Cole died in 1848, Knickerbocker William Cullen Bryant delivered the eulogy, recalling Cole's paintings "which carried the eye over scenes of wild grandeur peculiar to our country, over our aerial mountain tops with their mighty growth of forest never touched by the axe, along the banks of streams never deprived by culture and in the depth of skies bright with the hues of our own climate; skies such as few but Cole could ever paint, and through the transparent abysses of which it seemed that you might send an arrow out of sight."[23]

Deeply moved by Bryant's testimonial, Jonathan Sturges, a patron of the arts, commissioned a painting by Asher Durand. Titled *Kindred Spirits*, it portrays Bryant, the poet, and Cole, the painter, on a rocky ledge in the Catskills wilderness. The painting is a lasting testament to the respect that these artists, writers and patrons held for one another.

As the artistic, literary, and mercantile capital of the nation, New York City was the artists' and writers' gathering place. At this time it was still a relatively small city and noted for its congeniality. In 1827, a group of these "kindred spirits" founded the Sketch Club, the ancestor of the Century Association.

Membership in the Club was by invitation only and included, in addition to leading artists and writers, their patrons as well as a few amateurs who shared an interest in the fine arts. It was important not so much for its influence on the Highlands as for the opportunity it provided for artists and writers to come together, an association which would later produce collaborative efforts featuring the Highlands in new books about scenery.

The club permitted a cross-fertilization of ideas and was organized to provide both a convivial atmosphere and an opportunity for productive work. Originally, the meetings were devoted to impromptu drawing and writing. The host for the evening would provide a subject, usually from literature, and the artists and writers would spend an hour on it. At the end of the evening, the host would collect the sketches and keep them. As time went on, writers and artists would sometimes have different subjects, and eventually drawing was abandoned altogether. Occasionally communal authorship was tried, whereby three members would compose a single poem on topics such as "The Sublime" or "Character."

Although members took themselves seriously, they used the gathering for fun and entertainment. The group met on Friday evenings, in an aura of secrecy. The Club had no official address; its members met in each other's homes. Club meetings would be announced crypticly in newspapers under the initials of the host for the evening. The January 14, 1830 announcement in the *Evening Post* was typical: "S.C.; H.I. 49 Vesey St." To members it was clear that this meant that Henry Inman was entertaining; however, to the *Post* readership it was intriguingly provocative.

William Emerson, the club's secretary, described the atmosphere of the meetings as a "peaceful, social, laughing, chatting hubbub." Amid discussions of politics and the economy were interspersed speeches on such matters as the "domestic economy . . . of Bull-frogs," and the "combustion of pea nut shells," which was explained by Thomas Cole. The Club also watched demonstrations of phrenology, performed magic tricks, and on one evening, at the home of Robert Weir, raised a ghost. Mock punishments were also meted out, as in the case of William Cullen Bryant, who was instructed to write and publish an account of an art exhibition he had not yet seen. The article appeared, unsigned, in the *Evening Post* the next day.[24]

Poems like the "Culprit Fay" and tales such as *Knickerbocker History* clearly sprung from this funloving tradition which was cultivated at the Sketch Club, and the Highlands, so close to the literary capital, were familiar to all of the club's members and soon began to be featured in new works. As friends and members of an elite group, the club members promoted each other's work and collaborated on a number of books on American scenery which were bought by an eager public. The *Homebook of the Picturesque*, published by George P. Putnam in 1852, contained engraved prints based on paintings by artists such as Thomas Cole, Asher Durand, Robert Weir, Jasper Cropsey, and Frederic Church. These were accompanied by vignettes written by Knickerbockers James Fenimore Cooper, William Cullen Bryant, Nathaniel Parker Willis, and others.

Selling at $7.00 clothbound, the book was a popular gift item. *Picturesque America* was another such publication which contributed to the mutual fame of the writers and artists. Edited by William Cullen Bryant, it contained hundreds of illustrations by Hudson River School painters. Both books featured the Highlands and promoted the region's reputation abroad.

Interest in such books lasted for about fifty years and fueled nationalist sentiment about America's wilderness, scenery, and natural resources. The books helped to define aesthetic values for the public and allowed even the armchair traveler to become familiar with America's natural wonders. The most ambitious work of this type was *American Scenery*, a two-volume set written by N. P. Willis in 1840. Its 119 engravings are each accompanied by a legend, poem, anecdote, or essay which adds romantic association to the scene. Although it was illustrated by watercolorist William Henry Bartlett, an Englishman and not a Hudson River School Painter, it cannot be overlooked among scenery books.

Typically, these volumes began with a statement defending the American

W. Bartlett, "Indian Falls," 1840 engraving from *American Scenery* by N.P. Willis. Originally published by George Virtue, London and R. Martin & Co., N.Y. 1840.

W. J. Bennett, *West Point from Philipstown*, 1831 engraving. Courtesy of the New-York Historical Society.

landscape. "Certainly, our forests," said the author of *The Scenery of the United States Illustrated,* "fresh as it were, from the hands of the Creator, are, beyond dispute, incomparable." Elias Lyman Magoon, the author of *The Homebook of the Picturesque* (1852) wrote in his lead essay "that there are yet wild spots and wildernesses left . . . whence thought may take the wildest range." Such places, he believed "have ever developed the strongest patriotism."[25]

Without exception these books on scenery gave the Hudson and the Highlands a featured spot. The frontispiece of *American Scenery* is a view of Storm King and Pollepel Island, and in the opening pages Willis states: "of the river scenery of America, the Hudson at West Point is doubtless the boldest and most beautiful"[26] Ten of the 119 engravings are of the Highlands. Several others show Hudson Valley scenes, while the rest portray views of Niagara Falls, the Catskills and Lake Winnepesaukee as well as other areas of the country.

The engravings published in these scenery books found their way into the decorative arts. They were used as patterns for Staffordshire pottery and in British blue ware and pink ware, produced in great quantities beginning in the 1820's. In 1834 Frenchman Jean Zuber printed a popular set of scenic wallpapers which he exported from his headquarters in Alsace to the United States. The set, called "American Scenery," was based on engravings of the 1820s and was printed from wooden blocks. It showed views of nineteenth-century America particularly admired by Europeans: Niagara Falls, Virginia's Natural Bridge, Boston harbor, New York Bay and the Highlands at West Point.

In 1961, when Jacqueline Kennedy remodeled the oval Diplomatic Reception Room of the White House, she chose a set of Zuber wallpaper which had

Frontispiece of *American Scenery* published in 1840 by N.P. Willis with engravings by W. Bartlett. Originally published by George Virtue, London and R. Martin & Co., N.Y. 1840.

recently been removed from a mansion about to be torn down. The oval shape of the Diplomatic Reception Room seemed particularly suited to the panoramic sweep of the scenic views, and the scenes would be a welcoming addition to "the room people first see when they come to the White House," said Mrs. Kennedy. The antique paper was donated by the National Society of Interior Design, which announced that the same wallpaper was still being manufactured from the original wooden blocks.[27]

This international interest in American nature and scenery far outlasted the Knickerbockers. By mid-century their eminence had faded, although the painters enjoyed another twenty-five years of celebrity. The Hudson River School extended roughly from 1825—the date of Cole's first sketch trip—to 1875. The eclipse of the Highlands in painting began after 1860, when a second generation of artists discovered the monumental scenery of the Far West. Painters such as Thomas Moran and Albert Bierstadt painted large canvases of Yellowstone, the Grand Canyon, the Grand Tetons and Yosemite. Their work, sometimes known as the Rocky Mountain School, captured the drama of the western landscape. With the novelty of the these new, bold images of the Far West, American interest in the softer images of the Hudson Valley and the Northeast waned.

The public's interest in the Knickerbockers gave way in the 1850s to writers such as Emerson and Thoreau, who offered a transcendental view of nature, and to other writers who were preoccupied with new and entirely different issues. However, in the forty years that the Knickerbockers dominated the cultural scene, the Highlands emerged as a powerful and well-known literary image. Melville used it confidently as a metaphor in *Moby Dick*, in which he wrote:

> Human madness is oftentimes a cunning and most feline thing. When you think it fled, it may have but become transfigured into some still subtler form. Ahab's full lunacy subsided not, but contracted; like the unabated Hudson, when that noble Northman flows narrowly, but unfathomably through the Highlands gorge."[28]

Located near the New York art and literary capital, accessible by steamboat, and providing the backdrop for both Dutch folklore and Revolutionary War history, the Highlands served romantic artists and writers well as subject matter. Together they succeeded in establishing the Highlands as sacred ground, a "vast cathedral," in the words of N. P. Willis, "The Hudson a broad aisle, the Highlands a thunder-choir and gallery."[29]

While this romantic image was being popularized in New York and abroad, in the Highlands a period of intense industrial growth had begun, led by one of the druids' greatest allies, Gouverneur Kemble of "Cockloft Hall." Kemble moved to the Highlands in 1817 or 1818 where, in the words of his friend Washington Irving, he "turned Vulcan" and began "forging thunderbolts."[30] His new business, a gun foundry, established the Highlands as a center of manufacturing and heavy industry.

5. The Foundry at Cold Spring

A VISITOR to Cold Spring today would hardly guess that this picturesque mountain village across the river from West Point could have been the site of the largest and most modern iron foundry in the United States. But it was, with all the noise, smoke, and grime that entails. In one of the great contradictions of the nineteenth century, the Highlands region, which symbolized scenic beauty and brought humanity into close spiritual contact with nature, was at the same time at the center of an industrial movement which laid waste the forests of the Northeast, tamed vast expanses of wilderness, and turned generations of farmers into servants of the machine. For decades, as West Point imported worshippers to its shrines of history and scenery, Cold Spring exported pipes, cranks, gears, cotton presses, railroad engines, and cannonballs.

While the Revolutionary War enshrouded the Highlands with historic and romantic associations, it was the War of 1812 which set in motion the region's industrial development. Nearly a disaster, the war proved one thing to President Madison: heavy artillery was the key to modern warfare. He lost no time in establishing foundries to supply the government with guns and munitions. His decision determined the future of Cold Spring for the next century.

It is not hard to understand why Cold Spring was one of four sites in the nation selected for a federally subsidized foundry. Iron had been discovered in the Hudson Highlands, and conditions were perfect for its use. The hardwood

The West Point Foundry circa 1860. Courtesy of the Putnam County Historical Society. Rephotographed by Michael Spozarsky 1979, collection of Edward Rutsch.

forests of the area provided an abundant source of charcoal to fuel the furnaces. Mill streams plunging through the hills supplied water-power to drive the bellows. On the Hudson River, sloops and steamboats carried the products to market. And last but hardly least, the military academy at West Point turned out men who understood armaments.

In 1817, General Joseph Swift, a graduate of the Academy who later became commandant of the Post, approached Gouverneur Kemble with a proposal for the establishment of a foundry. Together, they raised $100,000 in capital from investors, and in 1818 incorporated the West Point Foundry Association with Kemble in charge. From the day that it opened until the end of the Civil War, the foundry expanded steadily.

The first contract with the government, in 1817, called for cannon, cannonade, round shot, grape shot, and "Turning and Chuggling Guns" to be supplied to the Navy Department. With government orders providing a steady flow of income, the foundry soon expanded its business to include box stoves, milling machinery, plowshares, bells, sash weights, and a variety of machine components such as shafts and cranks.[1]

Cold Spring benefited from its location on the Hudson close to the New York markets and the manufacturing centers along the river's shores. With the opening of the Erie Canal in 1825, the West Point Foundry was situated on the primary transportation route to the interior. It was in a unique position to supply the developing regions of the West. Later, when Pennsylvania coal replaced charcoal and the railroad replaced the sloop, the Cold Spring location proved to be a continuing advantage because of its excellent transportation linkages—the east-shore railroad stopped at Foundry Dock, and access to the coal fields was possible via the Delaware and Hudson Canal and the Erie Railroad spur at Piermont.

During the early years, the development of the steamboat, and later the railroad, created new markets for iron products. Kemble's firm won contracts for producing steamship engines as well as the first iron ship made in the country—the cutter *Spencer*. The engine for the first American-made locomotive, the *Best Friend* was also cast in Cold Spring, as were the locomotives *DeWitt Clinton*, *West Point*, *Phoenix*, and the record-setting *Experiment*, which in 1832 was capable of speeds up to eighty miles per hour. With the best available technicians, the foundry was in the forefront of its technology in innovation, production, and design.

Two economic depressions marked the first half of the nineteenth century, but Cold Spring grew and prospered, shaping the industrial revolution and being shaped by it. In addition to steam engines, boilers, and locomotives, the foundry supplied iron pipes to replace the wooden water systems of New York, Boston, and Chicago; sugar mills for the West Indies; cotton presses for the South; and decorative garden benches for an increasingly affluent population at home. An intact sugar mill produced in 1861 at Cold Spring is now preserved as a landmark at the Hacienda La Esperanza in Puerto Rico. However, it was government contracts for armaments which sustained the West Point Foundry

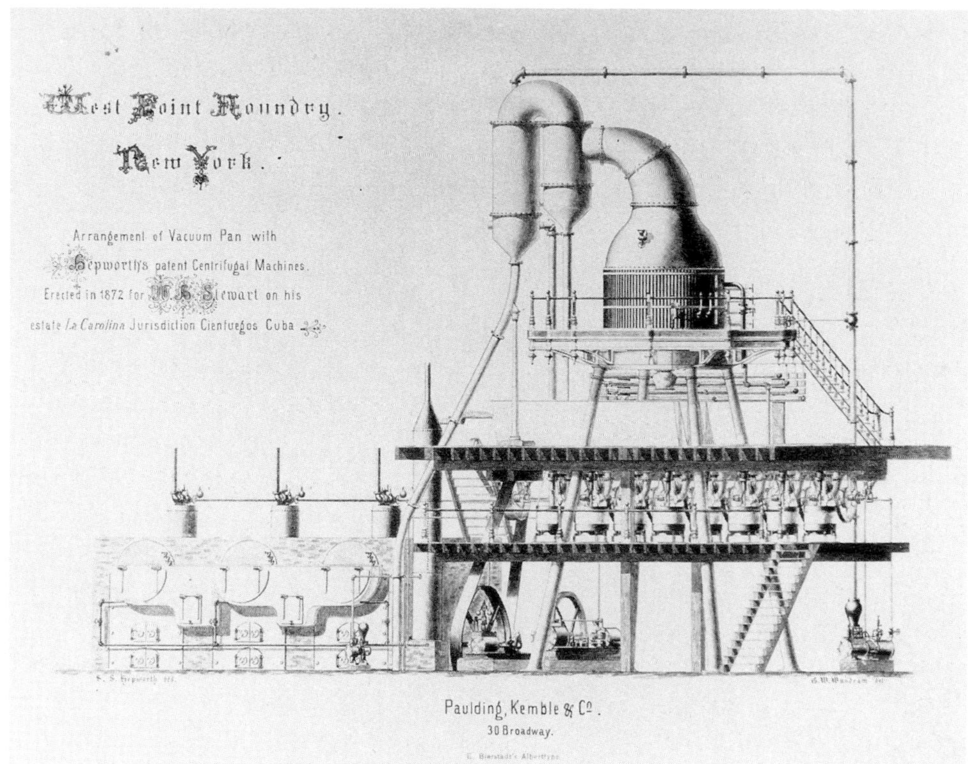

Sugar machinery produced by the West Point Foundry c. 1873. Courtesy of the Putnam County Historical Society. Rephotographed by Michael Spozarsky 1979, collection of Edward Rutsch.

at Cold Spring through fat times and lean and subsidized a continual process of modernization.

At first, the foundry employed some one hundred men. They mined iron, cut timber for charcoal, and hauled it by ox team to Cold Spring. But skilled mechanics needed for the foundry were not available. Kemble, who had studied technology in Europe, knew that the best mechanics were found in England, Scotland, and Ireland. Although strict emigration laws controlled the rights of skilled workmen to leave British shores, Kemble found ways to circumvent them. He was not above smuggling if necessary. M. Wilson, a Cold Spring resident during those years, recounts how it was done:

"Laboring men, but not mechanics could leave Europe. It required sharp practices for laboring men to leave. Consequently a company of laboring men was put on board a ship in the harbor of Belfast, Ireland, to sail to the United States, but on the eve of starting, mechanics were substituted for the laborers."[2] When the deception was discovered, a British war ship pursued the Cold Spring-bound mechanics but failed to catch up with them.

Once they arrived, Kemble provided housing for his workers, schools for their children, and eventually a Catholic Church, although he himself was a Presbyterian. The Chapel of Our Lady, constructed in 1828 on the shore of the Hudson in Cold Spring, is said to have been the first Catholic house of worship in the Hudson Valley north of Manhattan and aroused considerable public attention.

The dedication ceremony was reported in the *New York Mirror* and attended

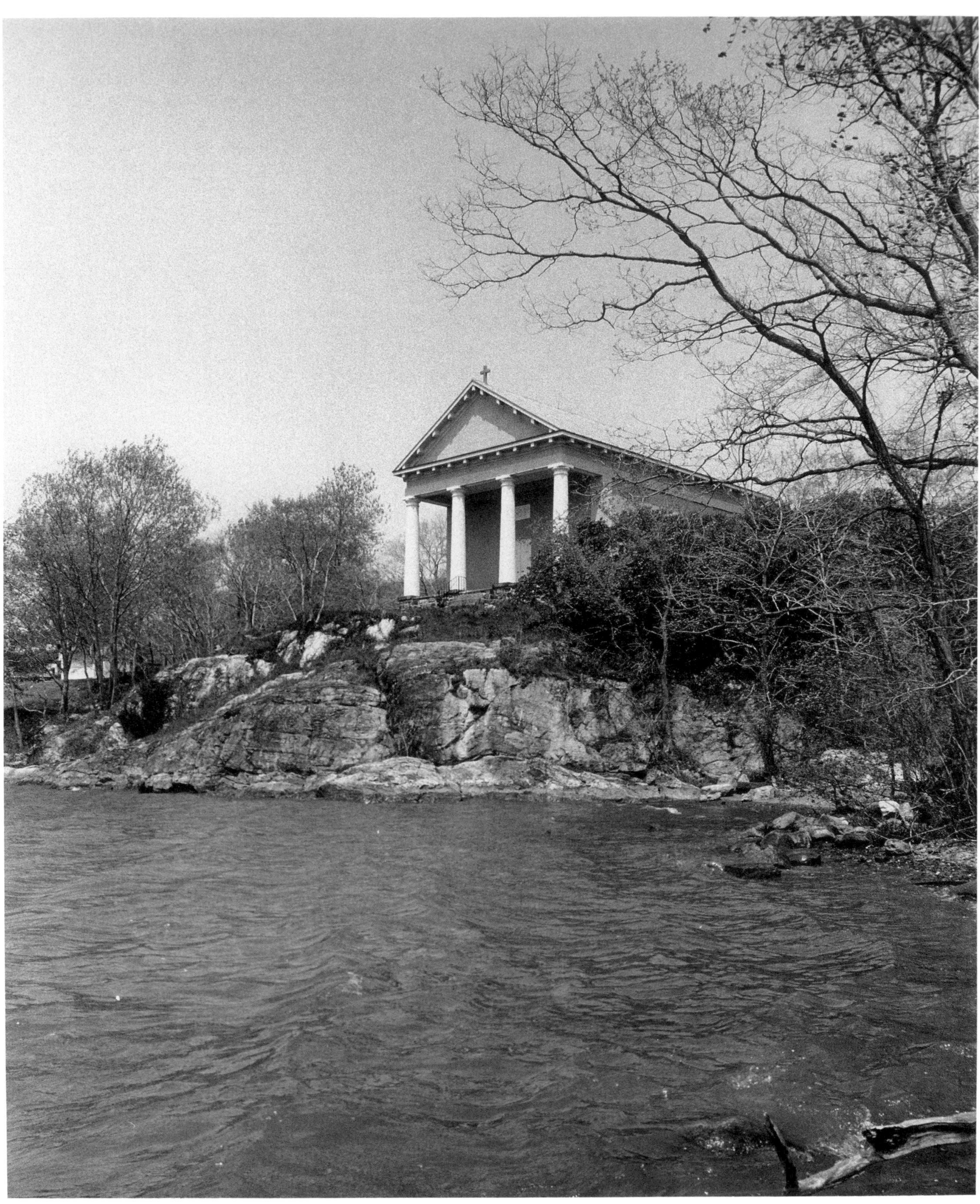

by Cold Spring residents as well as visitors from across the river and New York City. People arrived on foot, by rowboat, and by sail for the first solemn high mass. Later Robert Weir, the famous drawing professor at West Point, made an engraving of the Chapel, which contributed to its general acclaim. N. P. Willis included it as one of the 119 scenes in *American Scenery*. His accompanying text included the following passage: "This little chapel, so exquisitely situated on the bluff overlooking the river, reminds me of a hermit's oratory and cross which is perched similarly in the shelter of a cliff on the desolate coast of Sparta." He adds, irreverantly, that "it is a pity (picturesquely speaking) that the boatmen on the river are not Catholics; it would be so pretty to see them shorten sail off Our Lady of Cold Spring, and uncover for an Ave-Maria."[3]

The simple Greek revival style church was designed by sixteen-year-old Thomas Wharton, and remained in use until 1907 when a new, larger church was built on Fair Street. Falling into disrepair and badly burned in 1931, the chapel was purchased from the Catholic Church, restored and reopened in 1977 as a house of worship for all faiths. It is a monument to the Irish Catholic workers whose labor brought to the United States not only the foundry but also much of its industrial achievement.

Gouverneur Kemble, gentleman, bachelor, two-term Congressman, and President of the foundry was the village's most celebrated citizen. On December 13, 1852, Mrs. Elmira Southard wrote her mother describing the ceremony with which he was received on his return home from abroad:

> Mr. G. Kemble has arrived from Europe tonight. The Village has welcomed him with bonfires and illuminations. A procession met him at the Cars and escorted him across the street where the Kemble Guards received him. An address was spoken by Lawyer Dykeman and responded to by him. He then passed down the street among the Citizens, and to a carriage in front of our house, which was drawn by six White Horses. The Procession carried torches and we had fine music from the band. It was a fine Sight, I wish you could have seen it. The whole village joined in the welcome and honored his Grey Head. I have never seen such a turn out in Cold Spring.[4]

It was in 1836 in Washington that Kemble, who was then serving as Congressman, met Robert Parrott. He was impressed with the man and hired him to become superintendant of the foundry. The Highlands were already familiar to Parrott, who had started his military career at West Point, where he specialized in artillery. As "Assistant of Ordinance" in Washington, he also knew a great deal about modern weapons.

Joining Kemble's manufacturing venture, Parrott spent the next twenty years indulging his major interest—the development of artillery. To test the effectiveness of his new cannon, he shot at Crow's Nest and Storm King across the river. As steamboat passengers enjoyed the scenery of the Highland's venerable hills, Parrott and his men were busy blasting away at it.

But on the eve of the Civil War, Parrott had produced a cast-iron weapon of unparalled quality. It was a simple and efficient rifled cannon, which had a

Chapel of "Our Lady of Cold Spring."
Copyright © 1990 by Robert Beckhard.

c. 1860 portrait of Gouverneur Kemble. Courtesy of the Putnam County Historical Society. Rephotographed by Michael Spozarsky 1979, collection of Edward Rutsch.

longer range than any other of the period. According to Pelletreau's *History of Putnam County*, its unique construction consisted of a "cylinder made by coiling a bar of wrought iron, welding the coils together, and thus forming a cylinder which was turned and shrunk on the breech of the gun, thus preventing all danger of bursting."[5] In 1860, the foundry produced ten-pounders of this design, and increased the size in succeeding years to twenty- and thirty-pounders and ultimately 300- and 400-pounders. Technology, not strength, wins wars, and the "Parrott gun" is credited by many with turning the tide of the Civil War in favor of the Yankees. Its fame was so great that it firmly established the foundry's reputation for inventiveness. Jules Verne chose the West Point Foundry at Cold Spring as the place of manufacture for a projectile that was to journey into outer space in his "Trip to the Moon."[6]

All of Cold Spring benefited from the continuous growth of the foundry during its first sixty years. By the time of the Civil War, there were 1400 men working at the foundry, and they supported a high level of prosperity in the village. Doctors, lawyers, architects, carriage makers, druggists, and milliners are among those listed in the 1867 Beers Atlas as residents of Cold Spring and

Proving a 300-pounder Parrott gun on the grounds of the West Point Foundry. Courtesy of the Putnam County Historical Society. Rephotographed by Michael Spozarsky 1979, collection of Edward Rutsch.

neighboring Nelsonville. Blacksmiths, butchers, tailors, and grocers all set up shop to serve the needs of a growing village. Churches were built, offices, stores, inns, and homes sprang up. Carpenters did unprecedented business.

As Cold Spring prospered, it became a cultural center for artists, writers and the social elite, largely because of its proximity to West Point and the personal influence of Gouverneur Kemble, a patron of the arts, and of Hudson River School artists Robert Weir and Thomas Rossiter. Weir, an acclaimed landscape painter and professor of drawing at West Point, kept frequent company with the businessmen across the river. Rossiter made Cold Spring his home in 1860 in a house of his own design. His painting, *A Pic-Nic on the Hudson*, testifies to the close relationships between the officers of West Point, the members of Cold Spring society and the artists and writers who gathered there. It portrays Rossiter's friends and neighbors, including George Pope Morris, editor of the *New York Mirror*, West Point professor Robert Weir, and Gouverneur Kemble (see color plates).

Kemble's elegant home, "Marshmoor," was the hub of this social activity. Washington Irving, James Kirke Paulding and other Knickerbockers gathered often at the Cold Spring estate. The great Hudson River School artist, Asher Durand, visited there and painted the scene from Kemble's lawn in the mists of dawn.

On Saturday nights, dignitaries who were visiting West Point would ferry over to "Marshmoor" for the weekly open house, and when the prestigious Board of Visitors would come to the military academy for the annual examinations, Superintendent Thayer would send them across the river to enjoy the hospitality of the president of the foundry. There they dined on the fine foods prepared by Kemble's cook and drank the wines from his well-stocked cellar as they lingered over conversation. Thayer himself was a frequent visitor at those open houses, enjoying the company of Hudson Valley gentry such as President Martin Van Buren, as well as the other prominent men and women who were Kemble's guests.

Nearby was the summer home of George Pope Morris, who, with his friend and business partner N.P. Willis published the *New York Mirror*, a weekly magazine of literature and the fine arts. The *Mirror* printed works by Knickerbocker writers and engravings of Hudson River School paintings, as well as columns of commentary on European culture and articles about Hudson River scenery and how best to enjoy it. On the last page was the sheet music for new songs.

Morris was a great leader and contributed to public appreciation of cultural affairs. The *New York Mirror* and its successor, the *Home Journal*, were among the most widely read papers of their day. With its articles and engravings of Hudson River scenes, especially of West Point and the Highlands, it did a great deal to communicate the perceptions of artists and writers to a broad audience in this country and Europe. Morris himself was an amateur poet and some of his works were quite popular, "Woodsman Spare That Tree" being the best known. Morris's estate was well known to all in the art world, and his elegant home, visible from the deck of a steamboat became part of the picturesque landscape—painted, sketched and lauded in guidebooks. Generally the opening stanza of his poem "Ida" accompanied guidebook descriptions of his Undercliff, as in Wallace Bruce's *The Hudson*, which observes: "Standing on the piazza, we see directly in front of us Old Cro'Nest, and it was here that the poet wrote:

> Where Hudson's waves o'er silvery sands
> Winds through the hills afar,
> Old Cronest like a monarch stands,
> Crowned with a single star!"[7]

On nearby Constitution Island, the two Warner sisters, Susan and Anna, lived more reclusive lives but wrote novels with equal enthusiasm, adding to the reputation of the Highlands both as an artists colony and as a setting where nature could be experienced. From their island solitude, the two sisters published a series of best sellers which made them famous but failed to draw them from their hermitage. By day they supported themselves on fish they caught and berries they picked, while by night they wrote long tales which focused on life in

Thomas Chambers, *Undercliff Near Cold Spring, Home of George Pope Morris*, 1850 oil on canvas. Museum of Art, Rhode Island School of Design, Museum Works of Art Fund, Providence, Rhode Island.

the Highlands, religious themes, and subjects such as gardening. Susan Warner's *Wide Wide World* and *Hills of the Shatemuck* were immensely popular in their day. On Sundays the sisters also held religious services for the cadets at West Point and on their deaths were buried at the Point, the only civilians so honored.

All in all, the mid-nineteenth century was a time of prosperity and optimism for Cold Spring, supported by a solid economic base. But while the effects of the industrial boom on Cold Spring were positive, the impact on the surrounding countryside was not. Enormous fires were required to melt down the fifty tons of pig iron which were used for a single casting. To fuel the furnaces, trees from the surrounding hillsides were felled and the wood turned to charcoal at an astonishing rate. By mid-century, much of the forest cover of the Highlands had been removed for charcoal, cordwood and ship timber.[8] In fact, without the discovery of Pennsylvania coal, the fires at the foundry might well have been extinguished, with contracts and orders unfilled.

Currier & Ives, *West Point Foundry, Cold Spring, N.Y.*, 1862 engraving. Typical of its time, this engraving of Cold Spring and the foundry is more romantic than industrial. Courtesy of the New-York Historical Society.

In addition to the deforestation, from Fort Putnam above West Point the signs of industrial activity at Cold Spring must have been hard to ignore. The smoke from charcoal kilns, the roar of the blast furnaces, and the ring of the mammoth trip hammer, were constant reminders of the foundry's industrial presence. In sheer size, the works were vast, consisting of "a moulding house; a gun foundry; three cupolas and three air furnaces; two boring mills; three blacksmith shops; a trip hammer weighing eight tons for heavy, wrought-iron work; a turning shop; a boiler shop; and several other buildings used for various purposes." During the peak of activity, in the mid-nineteenth century, it employed 1400 men.[9]

It is an interesting commentary on the nineteenth-century Romantics that, with rare exception, writers, poets and artists omitted the foundry from their descriptions and paintings of the Highlands landscape. The views of Cold Spring which appeared in Hudson River School paintings and guidebooks focused instead on the elegance and natural beauty which characterized the other side of village life. Indian Brook Falls, the view from Kemble's lawn, panoramas of the Storm King gorge, Morris' elegant home on the hillside, or the picturesque Chapel of Our Lady were appropriate subjects for prose and paintings, but the smoke and physical presence of the foundry were obscured in a shroud of mist or cloud. Mountain slopes appeared cloaked in lush vegetation, dark with the shadows of ancient trees which at the time could only have been imagined, since the forests of the Highlands had been repeatedly clear-cut. The foundry does appear in a scene painted by David Johnson around 1870 and titled *Foundry at Cold Spring*, yet the foundry is so dwarfed by the surrounding scenery, that if the painting were untitled, the foundry would appear to be a barn, an interpretation enhanced by cows resting in the foreground.

One of the few writers who remarked on the foundry was Fanny Kemble, the British actress, who was given a tour of the iron works by her American host. But far from lamenting its intrusion on the scenery, she finds sublimity in the "dark abodes and their smouldering fires, and strange powerful-looking instruments."[10] Another was Benson Lossing whose book, *The Hudson from Wilderness to the Sea*, written in 1866, when the foundry was at its peak of activity, devotes 70 of its 460 pages to describing the beauty of the Highlands. Lossing confines his description of the foundry to a single sentence, remarking on the sound of "deep breathing furnaces, and the sullen, monotonous pulsation of trip-hammers"[11] with a footnote description of the facilities. In one of the few other paintings showing the foundry, one by Victor de Grailly, this same sublime image appears. The foundry is seen from across the river as a small fiery glow, diminished by an imposing mountain scene.

It was not until the Civil War era that romantic artists recognized industrial America in their paintings. When they did, they again applied the concept of the sublime, using images similar to those of Kemble and Lossing. The sparks, smoke, and power of industrial machinery had the same capacity to inspire awe and terror as sublime landscape. This, coupled with emerging pride in industrial progress, was reflected in one of the more notable nineteenth-century American paintings of an industrial scene, a 48 by 63-inch canvas titled *The Gun Foundry* painted in 1866 by John Ferguson Weir, the son of Robert Weir. (see color plates)

John F. Weir grew up at West Point and was a frequent visitor with his family at the home of Gouverneur Kemble in Cold Spring. Trained as an artist by his father, he used Kemble's ironworks as the subject for a series of sketches from 1861–1866. These were ultimately pulled together in one canvas, which won him lasting critical acclaim, and captured the industrial image of the Highlands in a major work of art.[12]

The scene, which depicts the casting of a Parrott rifle, contains many elements of sublime aesthetics. From among the cavernous shadows of the forge, the molten iron casts a powerful glow on the human figures toiling beneath the huge crane and cauldron. At a distance, a group of visitors stand in awe.

Perhaps the close relationship of artists and writers with their industrial patrons colored their vision of the Highlands scenery. However the oblivion of the Romantics to Cold Spring industrialism in the decades leading up to the Civil War also reflected the religious, philosophical, and scientific views of the day, when nature and spirituality were one. The "noise of civilization and of industry" were threats to unspoiled nature and did not fit the image of the Highlands as hallowed ground.[13] America, while industrializing at a rapid pace, could not reconcile these conflicts.

Undoubtedly, the affiliation of the artists and industrialists also helped to foster the image of sublimity which marked the eventual recognition of industrial America in art. By the time Weir painted his famous *Gun Foundry*, industry was beginning to symbolize a new progressive vision, and the foundry at Cold Spring was peaking in productivity. However it soon began a long decline. Because of this, the romantic image of the Highlands persisted.

After its Civil War heyday the foundry continued to operate, but never again at wartime levels. When steel replaced iron, its fate was sealed. It closed in 1911. Today, little remains of the iron works, although shops and homes in Cold Spring still show the signs of a once-thriving industrial river village of the nineteenth century, the social center for artists, writers, and the industrial, political, and military elite.

6. Mountain Spa

IN MANY WAYS, the next chapter in the history of the Highlands parallels that of the Catskills and the Adirondacks of upstate New York. All three mountain regions were thought to posess therapeutic powers, particularly in the treatment of tuberculosis and the prevention of diseases such as cholera. At the turn of the century, this image of a healthful mountain countryside would contribute to the preservation of these areas as parkland. In 1850, however, the Highlands were just beginning to develop a reputation as a refuge from the epidemics which ravaged New York repeatedly throughout the nineteenth century. This occurred at the same time when the foundry was expanding and the Hudson River School artists were enjoying their greatest popularity. It lasted until about 1910, when health-seekers ventured to other spots more distant from the city.

Cholera struck New York City in 1832, 1848, and 1854, following outbreaks of yellow fever and malaria which had arrived from Africa with the slaves. In addition, pulmonary consumption, or tuberculosis, was so widespread and so devastating in its toll on human life that it was known as the White Plague. The death rate rose steadily in New York, where water supplies were of dubious quality and sewage treatment was inadequate, until by the 1890s it had surpassed that of Old World cities. The term "lung blocks" arose to describe those New York slums with the highest death rates from tuberculosis. These diseases

Sheep pasture on Storm King Mountain in Cornwall. Copyright © 1990 by Robert Beckhard.

did not discriminate; they claimed young and old, rich and poor, famous and obscure, precipitating a health crisis of major proportions.

Scientists discovered the causes of these diseases at the end of the nineteenth century, but until that time medical theory held that malaria ("bad air" in Italian), as well as yellow fever and cholera, were caused by a gas called "miasma" produced by organic matter decomposing under humid conditions. This bad air was thought to be particularly prevalent in swamps and salt water. In contrast, the fresh pine-scented air of high elevations appeared particularly healthful. Medical journals published papers supporting this theory. A typical one was "Evergreen Forests as a Therapeutic Agent in Pulmonary Phthisis" by Dr. Alfred Loomis, a prominent physician specializing in lung disease. Another physician wrote in the *New York Medical Journal* that "the turpentine exhaled from these pine forests possesses, to a greater degree than all other bodies, the property of converting the oxygen of the air into ozone, and, as this latter destroys organic matter, the air of such forests must be very pure, and consequently conducive to respiration."[1]

To escape the toxic effects of miasma, doctors recommended that their patients travel to the mountains. For residents of New York City, the Hudson Highlands offered the "nearest mountain air, which is completely separated from the sea-board," being north of what one writer described as a "death-line" on the Hudson:

> At the meeting of salt water and fresh, there is an *infusorial cemetery* . . . the myriads of insects which belong to each realm of the element dying with the touch of the other, and precipitating at once to the bottom—thus producing the twenty-five per cent. of animal remains found in the mud of all Deltas at the mouth of rivers. Whereabouts is this death-line on the Hudson? The water is brackish even as high as West Point; but there must be a broad margin, a mile or two in extent, where the full tide of the sea meets the perpetual down-flow of the stream—an insect "valley of the shadow of death" hitherto unrequiemed and un-named.[2]

In the Highlands, the fresh air of evergreen forests and the absence of miasmatic gases were thought to offer the conditions for good health. With the so-called "death-line" just a few miles to the south, the Highlands were the closest place to New York where patients could hope to recover.

Actual experience seemed to support the miasmatic theory. In 1850, Dr. Robert Southgate, an Army doctor at West Point, published an article in the *New York Journal of Medicine* comparing the effects of the cholera epidemics at the Military Academy to other locales. He reported that West Point had escaped the 1832 epidemic and felt only minor effects of the 1848 epidemic as a result of its distance from "marsh miasmata" and the beneficial effects of its mountain air. Elsewhere, others were reporting tuberculosis cures during vacations in the mountains.

On the basis of reports such as these, a New York physician named Gray began recommending that his ailing patients seek a cure in the Hudson Highlands. One of those who took his advice was Knickerbocker Nathaniel Parker Willis. Failing health had sent Willis on trips to Bermuda and the West Indies, and on instructions from Dr. Gray he spent the summer of 1851 at the Sutherland House in Cornwall in the hope that the mountain air would cure the tuberculosis he suffered. There, he began a slow process of recovery, and in 1853 he purchased property in Cornwall, setting up a country seat which he called "Idlewild." Through his weekly letters to the *Home Journal*, which were later compiled into a book, *Outdoors at Idlewild*, he began to publicize the healthful surroundings of the area he described as the "Highland Terrace":

> To many the most essential charm of Highland Terrace, however (as a rural residence in connection with life in New York), will be the fact that it is the *nearest accessible point of complete inland climate*. Medical science tells us that nothing is more salutary than change from the seaboard to the interior, or from the interior to the seaboard; and between these two climates the ridge of mountains at West Point is the first effectual separation.[3]

The move to Cornwall prolonged Willis's life by another fourteen years, a fact not lost on his readers. He was a popular man, and his articles were read by a large number of people. In his time, Willis ranked just behind Cooper and Irving as the most famous American man of letters abroad, and was known not only for his literary prowess but also as a colorful figure on the Broadway scene. Oliver Wendell Holmes wrote of the twenty-five-year-old Willis that he was "young. . . . and already famous. . . . He was tall; his hair, of light brown color, waved in luxuriant abundance. . . . He was something between a remembrance of Count D'Orsay and an anticipation of Oscar Wilde. Lowell described him as 'the topmost bright bubble on the wave of the town.' "[4] Willis carried on a lively correspondence with some of his readers, and when he spoke, people listened. In the December 1853 issue of the *Home Journal* he quoted the following letter which he had received from a reader:

> Mr Willis.—*Dear Sir*:—
>
> . . . I followed you to Idlewild with much interest, having a fellow-feeling on one point, at least, and watched to see whether you could get the mastery of disease. In your last letter, you say that you are no longer to be classed among consumptives. Alas! I can't say as much for myself, I fear. And on reading your lines, I resolved to write to you, as a once fellow-invalid, and ask, *What has cured you*? The doctors advise me to go South and take cod-liver oil, but their prescriptions do me no good . . .
>
> . . . with true respect, A. D. G.

To this, Willis responded by detailing his philosophy and daily regimen:

> The patient who troubles himself least about his disease . . . , but who perseveringly *outvotes it* by the high condition of the *other parts* of his system is the likeliest to recover. . . . I, for one, came very near dying—not of my disease, but of what my doctors took for granted. . . . I went to the Tropics, as a last hope to cure a chronic cough and blood-raising, which had brought me to the borders of the grave. I found a climate in which it is hard to be unhappy about anything—charming to live at all—easy to die. . . . I reached home in July, thoroughly prostrated, and, in the opinion of one or two physicians, a hopeless case. . . . After an unflinching self-examination, I came to the conclusion that I was myself the careless and indolent neutralizer of the medicines which had failed to cure me—that one wrong morsel of food or one day's partially neglected exercise might put back a week's healing. . . . There was not a day of the succeeding winter, however cold or wet, in which I did not ride eight or ten miles on horseback. With five or six men, I was, for most of the remaining hours of the day, out-of-doors, laboring at the roads and clearings of my present home. . . .
>
> With all this—and looking like the ruddiest specimen of health in the country around about—I am still . . . troubled occasionally by my sleep-robber of a cough; and, in Boston, the other day, on breathing that essence of pepper and icicles which they call there "East Wind," I was seized with the old hemorrhage of the lungs and bled myself weak again. But I rallied immediately on returning to this Highland air, and am well once more—as well, that is to say, as is consistent with desirable nervous susceptibility. The kiss of the delicious South Wind of today . . . would be half lost upon the cheek of perfect health.

Willis went on to comment on the importance of horseback riding and urged his readers to avoid sweetmeats, pastry and gravies. He advised care and nurture of the skin, approved an occasional glass of wine, and cautioned that sleep and clothing require more attention than they are usually given. Posture, too, was important. Above all, temper and ambition should be prevented from souring. He concluded on the note he had emphasized many times in his letters—the "medicine of 'Out doors at Idlewild'—the mingled salubrity of the climate of mountain and river around us."[5]

The *Home Journal* had fifty thousand readers, and with a terrible health crisis in New York, any one who could afford to do so followed his example and came to the Highlands for the cure.

Cornwall, in particular, benefitted from this attention. When Willis first stayed at the Sutherland House in 1851, there were only a handful of boarding houses in Cornwall. Twenty years later there were dozens of them. Summer boarders arrived in ever increasing numbers over a period of sixty years, the most prosperous time in Cornwall's history. Benson Lossing wrote that Willis' pen had been as potent as a magic wand in attracting summer residents from New York. Lewis Beach, who wrote a book promoting Cornwall at that time, remarked that once Willis moved to Idlewild, "victims of phthsisis" have flocked to the area by scores and hundreds for the "air treatment."[6]

The air treatment stressed rest, good food, exercise and year-round exposure to purifying influence of sun and unconfined air. A typical day consisted of croquet, salt baths, doses of cod liver oil, long walks through Cornwall's glens and along shaded drives, sleep and good food. For the seriously ill patient, pills, syrups and troches replaced daily exercise. Chilly or foggy weather meant confinement to the hotel room.

The mountain cure cost $12–15 per week, and the boarding season lasted from May 1 to November 1. For those who were ill but not bedridden, a visit to Cornwall also offered opportunites for exercise, relaxation, and trips to West Point. Willis wrote that "For enjoyment of military shows and music—for all manner of pleasure excursions by land and water, to glens and mountain-tops, fishing, hunting, and studying of the picturesque—Highland Terrace will probably be a centre of attraction quite unequalled."[7]

The Hudson by Daylight, a guidebook promoting the tourist qualities of the river valley, states that Cornwall attracted 6000 summer visitors in 1873. The book notes the advantages the boarding houses had to offer, including their proximity to a Western Union Telegraph office, a post office, and a circulating library containing more than 2000 volumes.

Other qualities which brought the summer visitor to Cornwall, in addition to the salubrity of the air, were the romantic appeal of the Highland scenery and the accessibility of the steamboat and the railroad. "There is no pleasanter place for New York families," Beach writes, "as a gentleman can easily reach New York and return the same day, with a delightful sail on the river. . . . It is in fact a sort of rural Saratoga and in the Summer Season is full of life and gaiety."[8]

The availability of fresh meat, milk, and produce also enhanced the Highlands' therapeutic image. Located at the northern gate of the Highlands, Cornwall bordered the farmland of the mid-Hudson Valley, producing high-quality wheat and dairy products as well as raspberries, strawberries, and apples. On the flanks of Storm King, sheep were pastured. Although the opening of the Erie Canal in 1825 led to a decline in agricultural production in the Hudson Valley and in the Northeast generally (since the abundant Midwestern grains could

House in Cornwall once used to accommodate summer boarders. Copyright © 1990 by Robert Beckhard.

now be cheaply transported east) farmers in Cornwall prospered, selling their goods to the boarding houses, where fresh produce was part of the health package so valued by city visitors.

The notion of healthful mountain countryside had many promoters and produced substantial economic benefits for the Highlands region. Cornwall residents wasted no time in taking advantage of the demand for housing for the summer boarders. Everyone who could opened up a spare room for the summer or drove to the lumber mill to purchase wood for an addition to the house. In the words of one observer, "the widow of modest means; the mechanic or laborer with a room or two to spare; the farmer in his old fashioned but comfortable home—all open their doors, during the season, to the city guest."[9]

According to Lewis Beach, boarding houses were numerous and varied, in the number of guests they received, the character of the accommodations and the prices charged. Several accommodated over a hundred guests, including the Glen-Ridge House, the Smith House, the Linden-Park House, the Mountain House, the Clark House, and the Lawrence House. The Glen-Ridge was the largest, accommodating about 350. Many of these houses can still be seen along the Angola Road below Storm King Mountain, although they are now private residences. They are recognizable by their size and their nineteenth century architecture.

Although physicians were the first to recommend the area, it was not long before the tourist industry capitalized on the positive publicity—not always in noble ways. For example, the railroad companies, which had just laid track in the Hudson Valley in 1848, used the fear of disease as a way to amass great profits. Companies propagandized the idea that the shoreline was a malarial swamp. The railroad was able to purchase riverfront properties at low cost and then resell them at a huge profit after making "sanitary improvements."[10]

Steamboat and railroad promoters spread the word about the Hudson River spas to their customers. Lewis Beach's book *Cornwall* describes the attractions for visitors: mineral springs, lakes and fishing holes, hiking routes up Storm King, scenic walks, and drives and places for boating and bathing. It provides the names of hotels and boarding houses, discusses the prices of land, and concludes with an appendix of railroad and steamboat schedules. Beach assures his readers that "We could give any number of well authenticated cases of consumptives relieved, and not a few of perfect cures."[11]

Such advertising had the desired effect, creating an economic boom both for Cornwall and for the transportation industry. As Willis reported, steamboats were not lacking for passengers:

> It is a gay sight . . . to see the "day boat" sweep up with twice as many inhabitants as the nearest village; crowds of city-dressed people, leaning over the balustrade, and the whole a gaily painted and confusedly fascinating spectacle of life and movement. Then the "evening boat," with her long line of lights, her ringing bells, and the magical glide with which she comes through the darkness, touches the wharf, and is gone; the perpetual succession of freight-boats; the equipages from the surrounding villages; and all the "run-

> ners," coachmen, porters, and "loafers," who abound upon the docks, swarming the bar-rooms in the intervals of arrivals, contibute to keep up an excitement.[12]

Among the steamers which made the three-hour run between New York and Cornwall were the *Mary Powell*, the *C-Vibbard* and *Daniel Drew*, the *Cornwall* and *Baldwin*, the *New Champion* and the *Jessie Hoyt*.

The works of local authors also contributed to widespread recognition of the therapeutic qualities of the Highlands. In *The Hills of the Shatemuc*, a best-seller by Susan Warner of Cold Spring, one of the characters sends his children to the Highlands for their health, because "he knows that this is just the finest country in the world, and the finest air, and he wants them to run over the hills and pick wild strawberries and drink country milk, and all that sort of thing."[13] The girls arrive, one of them pale and thin, and spend the summer fishing, picking berries, and learning about farming. In the process, they are restored to the pink of health.

Summer boarding was not confined to Cornwall nor were all the visitors ill. New concepts of health contributed to public interest in exercise and outdoor sports such as hunting, fishing, hiking, bathing, bicycle riding, and rowing. The out-of-door movement led to the development of summer camping and the institution of the summer vacation. Because of these shifting attitudes and the increase in the number of people with time and money to spend, the age of steamboat and railroad travel was accompanied by a surge of resort and recreation development to serve the new leisure class.

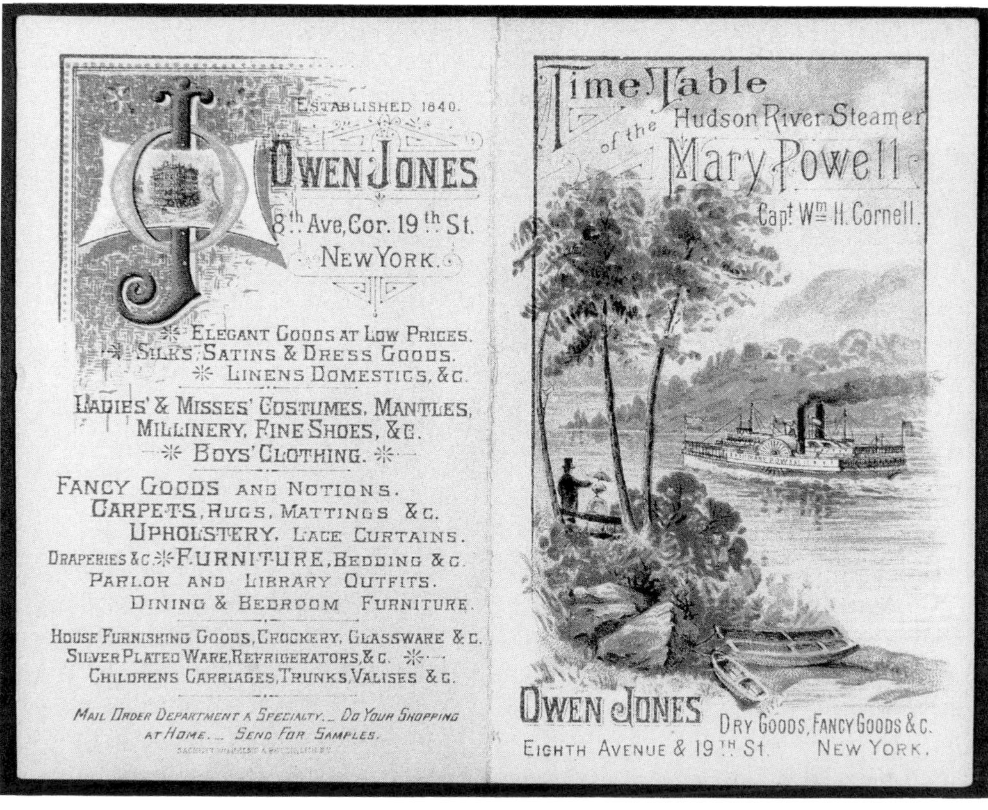

Timetable for the steamboat "Mary Powell." Courtesy of Constantine Sidamon-Eristoff.

The steamboat "Mary Powell."

Historians Raymond O'Brien and Roland Van Zandt have documented the transformation the mountain house resorts wrought upon the Hudson River Valley. There were dozens of them "multi-storied and built of wood, mansarded and flag-topped, usually on a promontory overlooking the Hudson."[14] The most famous resort was Catskill Mountain House but close behind it in popularity was Cozzens Hotel in the Highlands. Both were frequently portrayed in paintings and engravings and described in guidebooks. Cozzens' Hotel opened in 1849 on the grounds of West Point. When the original building burned down in 1861, the hotel then relocated to a promontory on the Hudson just south of the West Point Thayer Gate in adjacent Highland Falls, where it became part of the picturesque landscape. There it grew to be one of the largest summer hotels in the Hudson Valley and hosted such dignitaries as the Prince of Wales and served as the summer headquarters of West Point Commander Winfield Scott. Cozzens' mansion and adjoining cottages boarded as many as 500 guests in a season.

Many a wealthy New York businessman moved there for the summer bringing wife and children to this healthful spot and commuting to the city by ferry to Garrison and thence by rail to Wall Street. A frequent entry on Cozzens' guest list was J. Pierpont Morgan, who found Buttermilk Falls (as Highland Falls was known in those days) to be so lovely that he later purchased a large estate on the Hudson River bluffs nearby and became one of the area's more celebrated personages.

The hotel catered to the burgeoning outdoor movement and the continuing public interest in American scenery. Benson Lossing describes the attractions of Cozzens Hotel for nineteenth century visitors: "pleasure grounds of walks, drives and overlooks; paths winding through romantic dells and ravines and along clear mountain streams; opportunties for boating excursions to places of interest such as Buttermilk Falls, which could be explored from rude paths and bridges constucted so "visitors may view the great fall and the cascades above from many points."[15]

Roe's Hotel and the Parry House were other Highland Falls guest houses. On Iona Island, below Bear Mountain, the Grant mansion was converted to a hotel, and in Cornwall, there was the Storm King Mountain House. Cornwall and Highland Falls attracted different types of clientele. Summer visitors at West Point tended to be wealthier and spent their time in "dress and dissipation," dining on rich foods and attending parties and Hops. Many stayed for the entire summer season, from June to October. Cornwall drew a less well-to-do crowd, people who were more likely to live in boarding houses than in hotels and who stayed for weeks not months. Cornwall Society, according to Lewis Beach was made up of people "who prefer a more rational mode of enjoying themselves."[16] Cornwall residents, for their part, found the Highland Falls society to be haughty and condescending. The Reverend E. P. Roe, who lived in Cornwall and wrote popular books on gardening, held them in the utmost contempt:

> In fact, I have been led to believe that these gilded creatures are not aware of what Horace and Virgil and a host of other respectable people have said about gardening. Indeed, I am not sure that they are acquainted with the existence of those two worthy gentlemen named. Or they may indulge in Darwinian theory, and instead of going back to the first gardener for pedigree, hold that they are descended from sundry apes and oysters. From the mental, moral and physical developments sometime manifested, I should be at a loss to dispute their claims.[17]

Although the two towns attracted different clientele, both established reputations as fashionable resorts and health spas where the effects of miasma could be avoided. To a lesser extent, Cold Spring, across the river, also attracted boarders, while Iona Island was a favorite retreat for day visitors. It had a ferris wheel, carousel and picnic ground in addition to the pleasure grounds of its resort hotel and the vineyards of its former owner, Dr. Grant, who developed the Iona grape at this location.

More than any other person, Nathaniel Parker Willis was responsible for creating this reputation, and his column also did a great deal to maintain and enhance the romantic appeal of the Highlands through the latter half of the nineteenth century. It was in his letters to the *Home Journal* that he first advanced the idea of changing the names of some of the more scenic mountains, rivers, and villages. "In the varied scenery of our country," he wrote," there is many a natural beauty, destined to be the theme of our national poetry, which is desecrated with any vile name given it by vulgar chance." Willis began dreaming up new names, based on Indian legends or poetic images, and proposed them in his column. In the vicinity of his estate at Idlewild, he found several places to have offensive names, such as Murderer's Creek. Others, such as Bull Hill, he considered to be uninspiring descriptions of the grand scenery they identified. So he began to refer to them with new names in the weekly columns, and it was not long before the public and the postmaster joined in. Under his persuasion, Murderer's Creek became Moodna Creek, a name it bears today, while Bull Hill became Mount Taurus. After a "summer of discussion" residents of an unnamed hamlet near Idlewild drew up a petition to the Postmaster-General urging the adoption of the name *Moodna* for their small village, and it was at this time that Willis successfully advocated changing the name of Butter Hill to Storm King.

His fascination with such names began with the purchase of his own property in Cornwall. He writes: "When I first fell in love with it, and thought of making a home amid its tangle of hemlocks, my first inquiry as to its price was met with the disparaging remark, that it was of little value—'*only an idle wild!*, of which nothing could ever be made.' And that description of it stuck captivatingly in my memory. 'Idle-wild!' 'Idle-wild!' "[18]

Willis' efforts were far from silly or trite. They accurately reflected the value placed on the scenery by the Romantics of the nineteenth century. In bestowing romantic names on the mountains which form the northern gate to the Highlands, Willis captured the emotional appeal of the Highlands scenery and helped assure its preservation a century later.

Idlewild, home of N.P. Willis.

The pursuit of the mountain cure amid inspiring romantic scenery lasted long after Willis' death and permanently changed the pattern of land development in the Highlands. By the 1880s rail transportation made the Catskills and the Adirondacks more accessible to the city population, and these took over as the more fashionable resorts and health spa regions. However, in the Highlands, after the Civil War, a new type of development was taking hold—estates and summer homes for wealthy New Yorkers inspired to imitate the European aristocracy. The value of river property rose, as increasing numbers of city dwellers set up country seats in the healthful surroundings of the Highlands, where they were close enough to commute to New York on the new Hudson Line of the railroad. The few farmers who had been trying to coax a crop from the Highlands' stony soil sold their property and moved inland, as Willis noted in the April 1853 issue of the *Home Journal*:

> I met one of my neighbors yesterday, seated in his wife's rocking-chair on top of a wagon-load of tools and kitchen utensils, and preceded by his boys, driving a troop of ten or fifteen cows. As he was one I had always chatted with, in passing, and had grown to value for his good sense and kindly character, I inquired into his movements with some interest. He was going (to use his own phrase) 'twenty miles farther back, where a man could afford to farm, at the price of the land.' His cornfields on the banks of the Hudson had risen in value, as probable sites for ornamental residences, and with the difference (between two hundred dollars the fancy acre, and sixty dollars the farming acre) in his pocket, he was transferring his labor and his associations to a new soil and neighborhood. With the market for his produce quite as handy by railroad, he was some four or five thousand dollars richer in capital, and only a loser in scenery and local attachments. A Yankee's pots and kettles will almost walk away on their own legs, with such inducement . . .
>
> . . . A *class who can afford to let the trees grow* is getting possession of the Hudson; and it is at least safe to rejoice in this, whatever one may preach as to the displacement of the laboring tiller of the soil by the luxurious idler. With the bare fields fast changing into wooded lawns, the rocky wastes into groves, the angular farm houses into shaded villas, and the naked uplands into waving forests, our great thoroughfare will soon be seen (as it has not been for many years) in something like its natural beauty. It takes very handsome men and mountains to look well bald.[19]

The transition from marginal farms to lush estates changed the character of the landscape in the Highlands in dramatic ways. It was well under way in 1853, when Willis wrote this commmentary, and continued until the turn of the century. During this time, the river villages of the Highlands also prospered through the activities of the brickyards, the nearby manufacturing mills and the Foundry at Cold Spring, which expanded significantly in the Civil War era. This "new class who can afford to let the trees grow," included both village residents and those with new country seats. They created a unique business opportunity for the Downing nursery in nearby Newburgh, for they required not just trees and shrubs, but advice on landscaping. For the nurseryman, a man who had spent his youth exploring the romantic scenery of Newburgh Bay and the Highlands, this new clientele offered a way to test the ideas he was developing on the relationship of a home and its grounds to their natural surroundings. The result was the birth of a new American art form: the landscape garden.

7. The Landscape Garden: The Landscapes and Architecture of A.J. Downing

On Saturday, March 14, 1835, an unsigned article appeared in the *New York Mirror*, titled "America Highland Scenery. Beacon Hill." The article urged lovers of the picturesque who had confined their summer touring to the Catskill Mountain House at Pine Orchard and Cozzens Hotel at West Point to discover the scenery from Mount Beacon, a "commanding" summit where beacon fires had been lighted during the Revolutionary War. The author advised *Mirror* readers that they would be rewarded with a view so sublime it would "infuse new blood in your veins." The sunset from Beacon Hill, the author said, surpassed the reknowned sunrise from the Catskill Mountain House, and the scenery was more cultivated and luxuriant than the bleak and savage countenance of the Catskills. The Hudson—a bare line in the distance from Pine Orchard—was a captivating feature of the landscape from the summit of Beacon Hill:

> The Hudson—the prince of rivers! It appears to you that you see it rising like a silvery rivulet, thirty miles north of you in those distant hills, gradually widening, meandering, spreading life and freshness around, until here it loses itself, a deep, broad, powerful torrent in the rent chasm of the Highlands . . .
>
> . . . In every direction the country is full of beauty, and presents a luxuriant and cultivated appearance which is rare in mountain views in America. . . .

> Neat farm-houses are profusely scattered over the green and fertile fields, and here and there along the river is seen the beautiful and costly villa. . . . But ah! the eye wanders to that glorious spectacle, the noble chain of hills which forms the boundary, the frame, the setting to this superb picture.[1]

The author of this essay was Andrew Jackson Downing, a nineteen-year-old from Newburgh, who loved to explore the countryside of Newburgh Bay and the neighboring Highlands. As he climbed and rambled, he spent a great deal of time thinking about scenic beauty, about the feeling of open spaces and about human dwellings as features in the natural environment. This essay in the *New York Mirror* was one of his early attempts to communicate his thoughts on the elements that make a landscape scenic: the visual effects of the natural boundaries between wilderness and cultivated countryside; "beautiful and costly villas"; and "neat farm-houses." Within a few years, he would refine these ideas, publishing a series of books which would change the face of America. Downing's books and essays would teach generations of his countrymen how to design and landscape their homes. He would persuade countless communities to plant trees along their barren streets and establish public parks where all classes of people could come for outdoor recreation and enjoyment of nature. Along with Knick-

Andrew Jackson Downing (1815–1852). From A.J. Downing, *Architecture of Country Houses* reproduced from the reprint by Dover publications.

erbocker poet William Cullen Bryant, Downing would be a leading advocate for the creation of New York's Central Park.

Downing was a daring innovator, and the Highlands' landscape inspired him with a grand vision for America. He was a contemporary of many of the Hudson River School painters and Knickerbocker writers, and, like them, he saw the natural landscape as a work of art. Just as the painters sometimes saw ways of improving the scenes they painted, adding a twisted tree here and a boulder there, Downing, a nurseryman, knew that the home environment could be made more beautiful through landscaping, and he set about defining his own new art form—the landscape garden.

Steeped in the romanticism of his day, he took the ideas of the sublime and picturesque represented by the Highlands and applied them to the design of homes and home gardens. He showed his countrymen countless ways to bring their domestic lives into closer contact with nature and capture the emotional power of scenery in their own backyards. Doing so, he argued, would free their spirits and contribute to the beautification of America.

Born in October 1815, Andrew Jackson Downing was the son of a wheelwright-turned-nurseryman. When he was seven years old, his father died, and his older brother Charles took charge of the family and assumed control of the Newburgh nursery. Poverty prevented Andrew from going to college, and at age sixteen he joined his brother in the management of the family business.

As a fledgling nurseryman, Andrew began to train himself in landscape design. His clients were wealthy New Yorkers who purchased plants, trees, and shrubs to landscape their fine country seats on the banks of the Hudson. Downing visited their houses often, offering suggestions on how to use his nursery plants in the estates' landscaping.

At one of these summer villas, Downing had the good fortune to meet a foreign nobleman who took an interest in him and introduced him to society. Baron de Liderer, an Austrian consul general, shared Downing's passion for botany and minerology, and the two became fast friends. As the friendship grew, Downing was welcomed into the company of the Baron and his neighbors. In their homes he encountered a grace and refinement which fascinated him throughout his lifetime. At the country seat of the wealthy Edward Armstrong, Downing met Mr. Charles Augustus Murray, an English gentleman, whose manner impressed him a great deal. Downing began schooling himself, developing a gentlemanly polish and shedding his rough, country manners. He was later described by his friend George William Curtis as a tall, dark-eyed, Spanish-looking gentleman who carried himself with quiet dignity and a certain aristocratic hauteur.

Soon after he met Charles Murray, Downing was introduced to another Englishman, landscape painter Raphael Hoyle and struck up a friendship. The two of them roamed the hills and valleys of Newburgh Bay and the Highlands—Hoyle sketching the scenery, Downing observing the trees and flowers, and both of them sharing their feelings on art, life and beauty. This relationship stirred Downing's artistic desires, and by age twenty he was consumed with the idea of

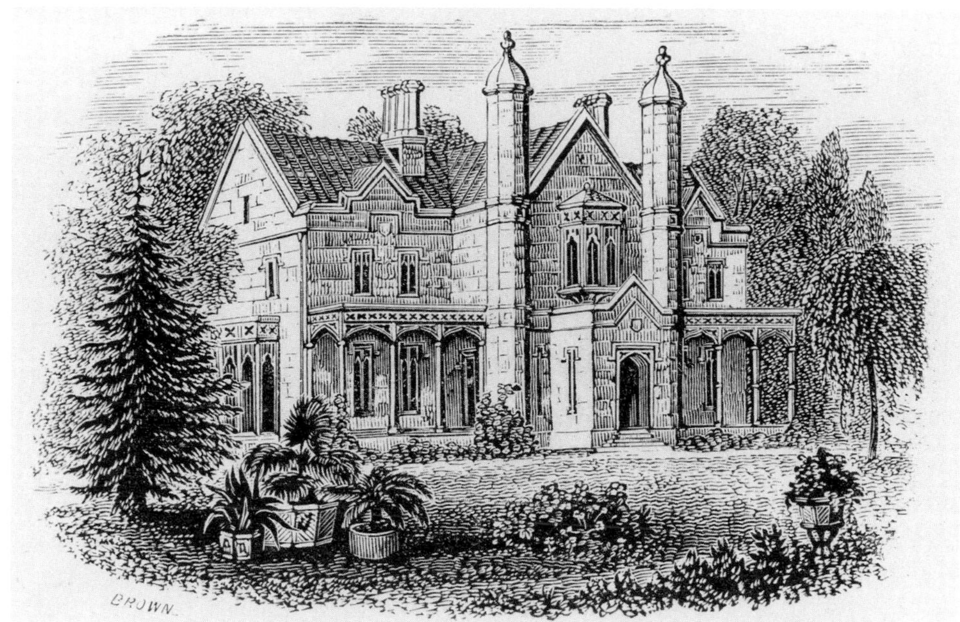

"Highland Gardens," home of A.J. Downing in Newburgh, N.Y. overlooking the Highlands. Reproduced from *Treatise on the Theory and Practice of Landscape Gardening*, by A.J. Downing. Courtesy of the Yale University Library.

designing a house and its grounds as a work of art equal to any painting or statue. To refine his thoughts, he extended his observations to the beautiful estates along the Hudson, where he was now warmly received.

Of all the estates on the Hudson, Downing most frequently visited "Locust Grove" at Fishkill Landing near Beacon Mountain. There he met and wooed the beautiful young heiress, Caroline de Windt, marrying her in 1838, when he was twenty-three. After his marriage, he bought out his brother's interest in the nursery, and immediately set about designing a Gothic villa for himself and his bride on six acres overlooking the Hudson in Newburgh.

The house, which he named "Highland Gardens," was Downing's first practical application of the principles of landscaping and architecture, which he had formed in years of studying Hudson River estates and river scenery. Derived from an English design, it was a turreted Elizabethan building surrounded by evergreens and ornamental shrubs that accented its picturesque setting. Downing added a large porch to give the residence an "American expression" and to unify building and landscape. His friend George William Curtis wrote of it:

> You fancied the estate extended to the river; yes, and probably owned the river as an ornament, and included the mountains beyond. . . . (T)he thicket seemed not only to conceal, but to annihilate, the town. At least, you felt that here was a man who knew that the best part of the landscape could not be owned, but belonged to every one who could appropriate it.[2]

So successful was this experiment that Downing began to set his thoughts on paper. He felt strongly that beautification of one's home and the surrounding land expressed the deepest patriotism, and he wanted to promote the landscaping ideas he had developed at Highland Gardens to a wider audience. In 1841, when he was twenty-six years old, he published his first book. The full title, which gives an indication of its scope and style, was *A Treatise on the Theory and Practice of Landscape Gardening, Adapted to North America with a View to the Improvement of Country Residences Comprising Historical Notices and General Principles of the Art, Directions for Laying Out Grounds and Arranging Plantations, the Description and Cultivation of Hardy Trees, Decorative Accompaniments of the House and Grounds, the Formation of Pieces of Artificial Water, Flower Gardens, etc. with Remarks on Rural Architecture*. The book soon ranked as a classic and became the definitive guide to landscape gardening for the next eighty years. It passed through ten editions, the last one published in 1921, and established Downing as the chief American authority on "rural art." According to U. P. Hedrick's *History of Horticulture in America*, landscape gardeners agree that the 1921 edition, updated by Frank Waugh, is still "without rival in any period or in any language in this field."[3] The success of the book catapulted Downing to international celebrity. Queen Anne of Denmark sent him a magnificent ring after she read it, and he was elected to the Royal Botanic Society of London and to the Horticultural Societies of Berlin and the Netherlands because of it.

The *Treatise* sets forth the principles of the art of landscape gardening which Downing developed in his discussions with artists, his hikes through the Highland scenery, and his visits to Hudson River estates. Instead of the formal, geometric, and highly stylized landscaping then in vogue, he promoted a fluid and graceful "natural" style. The natural style, he said, does not set itself apart from nature but uses the forms to be found in trees, surfaces of ground, and buildings in the surrounding landscape. Downing suggested that the natural style, which had originated in England, could be adapted to North America, improving the American scene and uplifting the spirits of the citizenry. Home beauty was an outward expression of inward good, he wrote, and for that reason possessed a moral and spiritual value. At the time Downing was writing, landscaping was confined primarily to the large landed estates. It was his belief, expressed in the *Treatise*, that landscaping could be applied equally well to small cottages. In beautifying their homes, Americans would be working in harmony with the fine natural scenery.

Along the banks of the Hudson River, where the constantly varying forms of water, shores, and hills offered a variety of home landscapes and distant views, the natural style was taking hold, Downing said. Hudson River property, he noted, particularly rewarded the efforts of the landscape gardener, because roads, walks, drives, and new plantations could be laid out to heighten the charms of nature and offer the best views of scenery. "A residence here of but a hundred acres, so fortunately are these disposed by nature, seems to . . . be a thousand in extent." By studying the examples offered by several of these properties, the principles of landscape gardening could be learned.

Downing's object in his *Treatise* was to provide advice to the hundreds of individuals wishing to ornament their grounds, but lacking a knowledge of how to proceed. Using the vocabulary of romanticism, he distinguished two types of gardens: the *beautiful* and the *picturesque*. The beautiful landscape garden, he said, was free and harmonious without display of power. It expressed unity, symmetry, and proportion. The picturesque one, on the other hand, was rude, violent, and irregular, frequently displaying power only, and expressing nature's struggle with opposing forces: the upheaval of mountains by convulsion, valleys broken by chasms.[4]

The expression of the beautiful and the picturesque involved different approaches, dictated by the architecture of the home. In Downing's view, the

Examples of the beautiful (top) and the picturesque (bottom) from Downing's *Treatise on the Theory and Practice of Landscape Gardening*. Courtesy of the Yale University Library.

beautiful was appropriate to a house of classical design—Italian, Tuscan, or Venetian—the effect being achieved through easy, flowing curves, soft surfaces, and rich, luxuriant growth. Trees should be smooth-barked varieties with rounded heads of foliage, such as ash or elm, planted apart, so their forms could develop to the fullest. Walkways should follow the natural lay of the land, and lawns should be mowed soft like velvet. Lakes should have masses of flowering shrubs on their borders.

By contrast, the house in a picturesque landscape setting should be an Old English or Gothic mansion or a Swiss cottage "with bold projections, deep shadows, and irregular outlines."[5] The grounds of the picturesque home should be characterized by spirited irregularity, abrupt and broken surfaces and growth "somewhat wild and bold in character." The picturesque was the form Downing preferred, perhaps in response to the wild scenery which had influenced his childhood. For this style he suggested massed plantings of rough-barked pines, larches and other trees of striking, irregular growth. Walks and roads should turn at sudden angles, preferably offering glimpses of distant views. Lawns should be mowed less frequently. Streams with bold shores and cascades in secluded dells were forms to be imitated and preserved.

The principle guiding Downing's work was the harmonious union of building and scenery. The country residence and its grounds should make "such a composition as a landscape painter would choose for his pencil."

Americans proved very sympathetic to this notion of the home landscape as a work of art and soon others began to promote it. A year after the publication of the *Treatise*, Edgar Allan Poe wrote a piece for the *Ladie's Companion* called "The Landscape Garden." This essay was later incorporated into the tale, "Domain of Arnheim," which Poe completed in 1847. The tale illustrates the fascination with landscape gardening sparked by Downing. The story is about a man who has inherited $450 million dollars and is faced with the decision of how to spend it. Should he indulge in the fashionable extravagances? Perhaps, instead, he should invest in political intrigue, or the purchase of nobility, or patronage of the arts and letters. Ultimately, the heir decides to use his fortune to create his own supreme work of art—a landscape garden. Wrote Poe: "it seemed to my friend that the creation of a landscape garden offered to the proper muse the most magnificent opportunities . . . the creation of novel moods of purely *physical* loveliness."[6] The place the heir chooses for this is Arnheim, a fictional spot described earlier, in images strongly recalling the Highlands, where Poe had been a West Point cadet.

In addition to its influence on art, architecture, and literature, Downing's *Treatise* also fostered tremendous public interest in new species of trees, shrubs, and flowers. Although Downing did a great deal to promote the use of indigenous species, he also discussed numerous "new discoveries" from around the world, which he sold at his nursery. The Downing Botanic Garden and Nursery offered such imported greenery as the Ginko, Weeping Cypress, Stinking Yew, Baffin's Bay Borealis, Patagonia Fitzroya, African Tamaris and Taurian Pine. These species, new to Americans, were brought to Newburgh by the Hudson

River whalers and traders.[7] Henry Winthrop Sargent of Beacon, who worked with Downing and issued supplemental editions of Downing's books after the designer's death, is said to have discovered in the wild, behind Mt. Beacon, the wonderful tree which perpetuates his name—the Sargent's, or Weeping Hemlock. It is believed that the proliferation of this romantic species throughout the Northeast is a result of Sargent's grafting.

Part of Downing's success can be attributed to social, geographic, and economic changes which shaped the Victorian era. Until Downing's day, there were few small homes with enough surrounding land to warrant elaborate treatment. Before that, there were city dwellings, homes in country villages and isolated working farms, with little in between. The colonial farm of his day was rarely landscaped at all; trees and shrubs were uncommon near a house.

In New York, industrial growth, expanding commercial activity, and resort development dramatically increased the number of people who could afford to build and maintain a suburban home. By mid-century, the construction of the railroad made commuting to the city from the surrounding countryside possible. More and more prominent businessmen built villas on the shores of the Hudson, and by 1853 "working men" were building "out-of-town houses" as well, according to *Harper's Magazine*.[8] The times were right for Downing's ideas

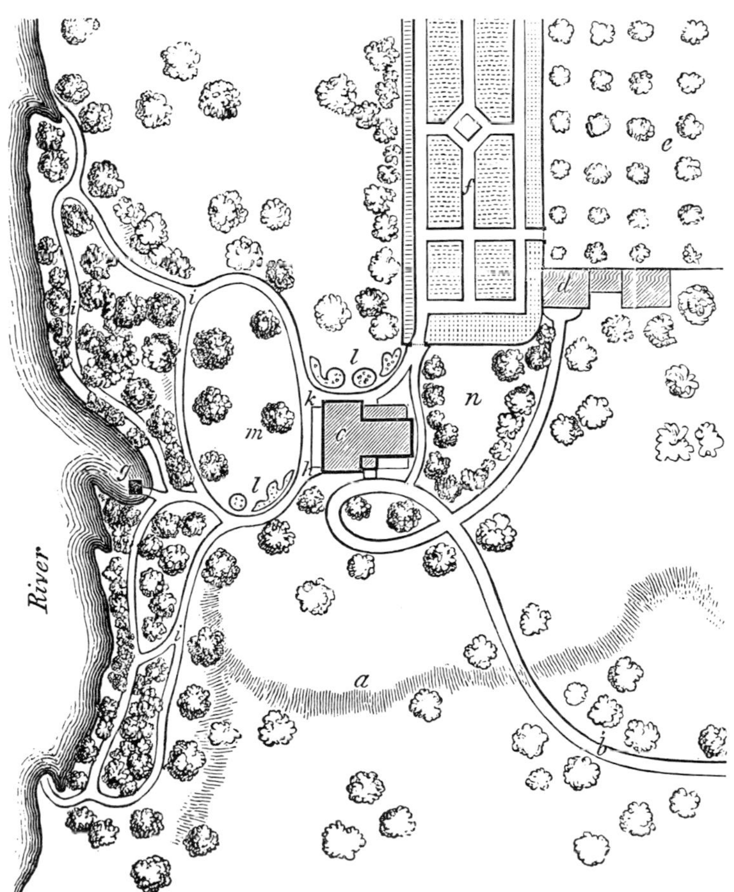

Design for the lay-out of the grounds for a riverfront property intended to complement a moderate-sized Bracketed Villa in the Italian style (see illustration p. 104) on about 150 acres of land. From A.J. Downing, *Cottage Residences*. Reproduced from the reprint by Dover Publications.

Bracketed Villa in the Italian Style from Downing's *Cottage Residences*. Although Downing designed many houses and landscapes himself, his greatest impact was as a popularizer of architectural and landscaping ideas. Through his illustrated books, Downing inspired a radical shift in taste among Americans, from the "tasteless temples" of Greek Revival to a new idea of simplicity. In the Hudson Valley, the Bracketed Style, identified by the decorative brackets under the eaves of an Italianate or Gothic villa, caught on like wildfire. The title of Edith Wharton's novel *Hudson River Bracketed* reflected this popular craze. Reproduced from the reprint by Dover Publications.

to be tried by Americans at all economic levels, and landscape gardening soon swept across the country.

A year after he published the *Treatise on Landscape Gardening*, Downing expanded his ideas from landscaping into architecture with a book titled *Cottage Residences*, which further established the pattern for the nineteenth-century suburban lifestyle. *Cottage Residences* offered designs for simpler, less expensive buildings and their grounds. Unlike earlier architectural books, it presented plans for the treatment of the entire lot, not just a design for the structure.[9] Downing showed the house in its setting, provided floor plans, elevations, and advice on structural problems; he discussed paint colors and suggested the layout of the grounds to provide for a kitchen garden and beautiful landscaping. He identified the varieties of trees to plant for the most complementary visual effects and suggested flowers and shrubs for a continuous succession of bloom and the sweetest fragrances.

Downing urged home builders to select settings which not only offered the best views but also enhanced the view in which the home itself was captured. He suggested that windows be oriented to improve the vista, and porches added to provide a transition from indoors to outdoors. To blend homes into their natural surroundings, Downing advocated the use of earth colors such as cream, fawn, drab, or stone grey instead of the then fashionable white paint. "*White* is a color which we think should never be used except upon buildings a good deal surrounded by trees, so as to prevent its glare." For the same reason he preferred the use of natural materials such as timber and stone.

Downing was also concerned with practical considerations. Utility and beauty were his two primary objectives, and he was far ahead of his time in suggesting that form should follow function in the convenient arrangement of rooms and the efficient arrangement of outbuildings, barns, fields and gardens. The importance of ventilation and drainage were first brought to public attention by Downing, and because of his writings, the kitchen, which had formerly been located in the basement of the house, was now moved to the first floor.

Downing recognized that the most inexpensive shape to build in was a square or rectangle, but he demonstrated that the inexpensive addition of bay windows, verandahs, and decorative trim would make a house look more irregular and richer in appearance.[10] Bay windows became a fad thanks to his writings, as did gables, turrets, buttresses, towers, ornamental roof slates and fanciful chimney tops. Decorative vergeboard—thick planking carved with Gothic motifs and often referred to as gingerbread for its lacy, frosting-like appearance on the eaves of a house—is another feature Downing promoted. All of these principles were extensively applied by homeowners everywhere, but especially in the Highlands, where his ideas originated. There the styles, colors settings and ornamental details he advocated were incorporated into nearly every nineteenth-century home—modest farmhouses and great estates alike. Today, these homes still predominate, and the Gothic effect, as Downing predicted, compliments the forms of the natural landscape.

Downing's ideal for a house in the Highlands was included as a "Design for a

Decorative verge board on a house in the Highlands. Copyright © 1990 by Robert Beckhard.

Design XXXII for a Lake or River Villa from A.J. Downing's *Architecture of Country Houses*. Of this design, Downing writes: "But let another person . . . live among such scenery as meets his eye daily in the Hudson Highlands, and he will often feel that a common-place, matter-of-fact, square house is an insult to the spirit of all that surrounds him. In such bold scenery, nature overpowers all and suggests all.

"It is in such picturesque scenery as this . . . that the high tower, the steep roof and the boldly varied outline, seem wholly in keeping with the landscape, because these forms in the building harmonize . . . with the pervading spirit of mysterious power and beauty in romantic scenery. . . . In composing this design, we frankly owe our indebtedness to the architecture of the Rhine towns for two features which distinguish it—the simple, square, high-roofed tower and the twisted column. The general composition of the design and the arrangement of the plan, of course, belong wholly to our own habits in domestic life, and have nothing to do with houses abroad." Reproduced from the reprint by Dover Publications.

River Villa" in his next book, *The Architecture of Country Houses*, published in 1850. The book also promoted his favored Italian, Venetian, Swiss, Rural Gothic and Bracketed styles and offered fourteen villa designs for the wealthy home-owner in addition to designs for cottages and farmhouses. One of these designs, a "Villa in the Pointed Style," was built in Garrison in 1867 for Edwards Pierrepont. In the same year, Downing decided to set up his own architectural firm and traveled to England where he met a young architect, Calvert Vaux, whom he persuaded to return with him to America to form a partnership. Although Vaux was only twenty-seven years old, he had already achieved distinction as a landscape painter and architect. His talents and philosophy complemented Downing's, and together they designed and constructed the houses and grounds of numerous estates along the Hudson and on Long Island.

Meanwhile, Downing began to expand his ideas from the landscaping and design of homes to developing landscaping plans for towns and villages, which at the time of his writing had no parks and so little greenery that the New England countryside was described by one observer as "a wild common over which the November winds swept with a pestilent force, with nothing to break them, except a pair of twin churches. . . . The toll-gate, the churches, the store lay strewn along a high-road three miles away . . . And yet so bare of trees was the interval that . . . I could see the twin churches, the tavern, and, with a glass, detect even a stray cow."[11]

In this barren countryside, there were pockets of beauty, and Downing used these to illustrate what could be done by the individual or town wishing to follow his theories. In an article on "Trees in Towns and Villages," published in a book titled *Rural Essays,* he suggested the creation of tree planting societies to facilitate this. In the 1840s and 1850s, following his advice, many towns in the

"Hurst-Pierrepont," in Garrison. Leading architects collaborated with Downing and contributed original designs. Alexander Jackson Davis, a highly successful New York City architect, provided the Gothic design for a *"Villa in the Pointed Style"* for Downing's *Architecture of Country Houses*. The plan was modified slightly in 1867 for a brick villa commissioned by Edwards Pierrepont for his estate in the Highlands. Copyright © 1990 by Robert Beckhard.

Highlands and throughout the Northeast began planting their streets with stately native elms, maples, oaks, ashes, tulip trees, and magnolias. Imported trees which had previously been the norm—the ailanthus and the poplar—were banished from the scene because of their foul odor and monotonous shapes. Ironically, these are now returning as popular street trees in cities because of their ability to withstand pollution. In the Highlands, mature maples line the road, many of them planted in Downing's day. Vast expanses of roadway are also forested.

One of Downing's greatest contributions was his advocacy for the creation of public parks in cities. His ideas on this subject were sparked by Mount Auburn cemetery, the 1831 creation of Dr. Jacob Bigelow. Dr. Bigelow broke with the tradition of the church graveyard and laid out an eighty-acre naturalistic landscape as a resting place for the dead. Located six miles from Boston, in Cambridge, it alternated lawns and monuments with masses of native trees on a site of undulating terrain. It was so popular as a place of quiet recreation for the living, that twenty years later, nearly every town and village had its rural cemetery. The cemetery in Nelsonville, adjacent to Cold Spring, dates back to this era and clearly shows Downing's influence on its design. At Laurel Hill Cemetery in Philadelphia, where records were kept, there were 30,000 visitors in 1848.

With such evidence of public interest, Downing argued in 1848, the taste for public parks, devoted solely to healthful enjoyment of nature, would spread rapidly. Three years later, he was advocating that New York City set aside 500 acres for a park to give its citizens breathing space and a recreation ground for rural refreshment. In this campaign, he joined poet William Cullen Bryant, who published editorials in the *Evening Post*, and wrote about the parks he had seen in England. Together they were responsible for the passage of the 1851 Park Act, which brought Central Park into being. In 1858, Downing's partner Calvert Vaux joined with Olmsted in submitting the winning design for the layout of the park. Downing's ideas about parks and nature would lay the foundation for the preservation of the Highlands decades later.

In 1851, Downing achieved the greatest honor of his lifetime when he was engaged by President Millard Fillmore to design and lay out the grounds for the Capitol, the White House, and the Smithsonian Institution. His plan called for a comprehensive landscape design of sylvan character with irregular meandering paths and wooded pools. He was never to see it realized.

In 1852, at the height of his fame, he boarded the steamboat *Henry Clay* with members of his family and friends bound for New York City. Twenty miles from their destination, however, the *Henry Clay* began racing with a rival boat *Armenia* and caught fire. Calm in the face of disaster, Downing assisted his friends and began throwing wooden chairs overboard to help those who had leaped into the water float to safety. Numerous passengers drowned or were burned, but Downing's efforts saved his wife Caroline and many others. At some point, Downing, who was an excellent swimmer, jumped into the water himself, but in coming to the rescue of a beautiful young widow, Matilda Wadsworth, he was drowned. During his last moments, with the widow and several others clinging to him, Downing uttered a prayer to God and slipped below the surface.

In tribute to his memory, friends erected a monument in Washington, D.C. It was in the shape of a vase which had decorated the lawn at "Highland Gardens" and bore the inscription: "He was born and lived and died upon the Hudson River." On the other side of the monument was a quote from one of Downing's essays: "The taste of an individual as well as that of a nation will be in direct proportion to the profound sensibility with which he perceives the beautiful in natural scenery."[12] This philosophy, that all Americans shared the

The rural cemetery in Nelsonville, adjacent to Cold Spring. Cemeteries such as this one were the precursors to public parks and were designed to be inviting landscapes. Copyright © 1990 by Robert Beckhard.

Thomas Chambers, *View of Hudson River at West Point,* nineteenth century oil on canvas. Courtesy of the Albany Institute of History and Art, Albany, N.Y.

Thomas Cole, *Storm King of the Hudson*, 1825 oil on canvas. Courtesy of Ball State University, Muncie, Indiana. This painting is thought to date from Cole's initial trip up the Hudson River. The painting was titled later, since in Cole's day Storm King was known as Butter Hill.

Sanford Gifford, *Storm King*, 1866 oil on canvas. Courtesy of the Fogg Art Museum, Harvard University, Cambridge, Massachusetts. Gift of Mrs. William Hayes Fogg.

John Frederick Kensett, *Hudson River Scene*, 1857 oil on canvas. This is the classic view from Fort Putnam painted by many Hudson River School artists. Metropolitan Museum of Art, New York, Gift of Mr. H.D. Babcock, in memory of S.D. Babcock, 1907.

William Joy, *Forcing the Hudson River Passage,* undated oil on canvas showing the 1777 Revolutionary War battle in the Highlands. Courtesy of the New-York Historical Society, New York City.

David Johnson, *A View from New Windsor, Hudson River*, 1869 oil on canvas. Collection of Elizabeth and Robert A. Sincerbeaux.

Victor De Grailly, (attributed), *Sailboats Off West Point,* undated oil on canvas. De Grailly, a Frenchman, never saw the Hudson River landscapes he painted. His works copied images from engravings by Bartlett and others and demonstrate the international popularity of American landscape in the nineteenth century. Courtesy of the Putnam County Historical Society, Cold Spring, N.Y.

Seth Eastman, *Hudson River With a Distant View of West Point*, 1834 oil on canvas. Courtesy of the Butler Institute of American Art, Youngstown, Ohio.

Jasper Cropsey, *Cold Spring on Hudson*, 1874 oil on canvas. Courtesy of the Newington Cropsey Foundation, Hastings-on-Hudson, N.Y.

Robert Walter Weir, *View of Hudson River*, 1864 oil on canvas. Courtesy of the West Point Museum Collections, United States Military Academy, West Point, New York.

Thomas Pritchard Rossiter, *A Pic-Nic on the Hudson*, 1863 oil on canvas showing friends and neighbors of the artist, who stands at the far left. Others include: Professor Robert Weir (on rock), George Pope Morris (white beard) and his wife (wearing a hat, to his left), Gouverneur Kemble (seated far right, without hat) and on the far right, Robert P. Parrott. Courtesy of the Butterfield Memorial Library, Cold Spring, N.Y.

John Ferguson Weir, *The Gun Foundry*, 1866 oil on canvas. Courtesy of the Putnam County Historical Society, Cold Spring, N.Y.

"Burning of the steamboat *Henry Clay*, Riverdale," July 28, 1852. From H. R. Murdock's *The Hudson River*, vol. 1. Courtesy of the New-York Historical Society.

landscape and that making it beautiful was something every patriotic individual should contribute to, was Downing's legacy.

It was this message which Downing had conveyed so successfully to his countrymen in the seventeen years since his article had appeared in the *New York Mirror*. By the time of his death, Downing's books on landscaping and architecture had begun to transform the American countryside. It was several decades before the notions Downing had advocated appeared in full flower; yet nowhere in New York State are his principles and philosophy of landscaping and architecture more evident than in the Highlands, where nearly every nineteenth-century style he popularized can be found. The population of the Highlands grew tremendously in the years when Downing's ideas held sway. It was also a time of relative prosperity, when many people could afford to build a home. Highlands residents clearly thought about the relationship of their homes to the landscape, echoing its forms, colors, lines and textures. Neil Larsen, a former New York State historic preservation officer in charge of the Hudson Valley region writes, here "the natural and the built environment have become as one. The tree-line is frequently interrupted by towers and rooftops; the shoreline by fortifications against the harsh elements. . . . The scenery is as much a part of the architecture as the architecture is part of the scenery. Everywhere the mountains are obvious to the eye. . . . Bold massive buildings set upon stone foundations reminiscent of cliffs, . . . [the] setting [is] carefully determined to both capture a view and be captured in one . . . the dramatic quality of the natural environment influenced a craftsmanship which is unsurpassed in any other place."[13]

Downing brought an appreciation of landscape—once the province of artists and writers—into the realm of the average homeowner. His ideas were embraced by all segments of society, including the most powerful business and political leaders. Some of them would take Downing's ideas to greater heights during the post-Civil War era, when railroad presidents reigned like kings.

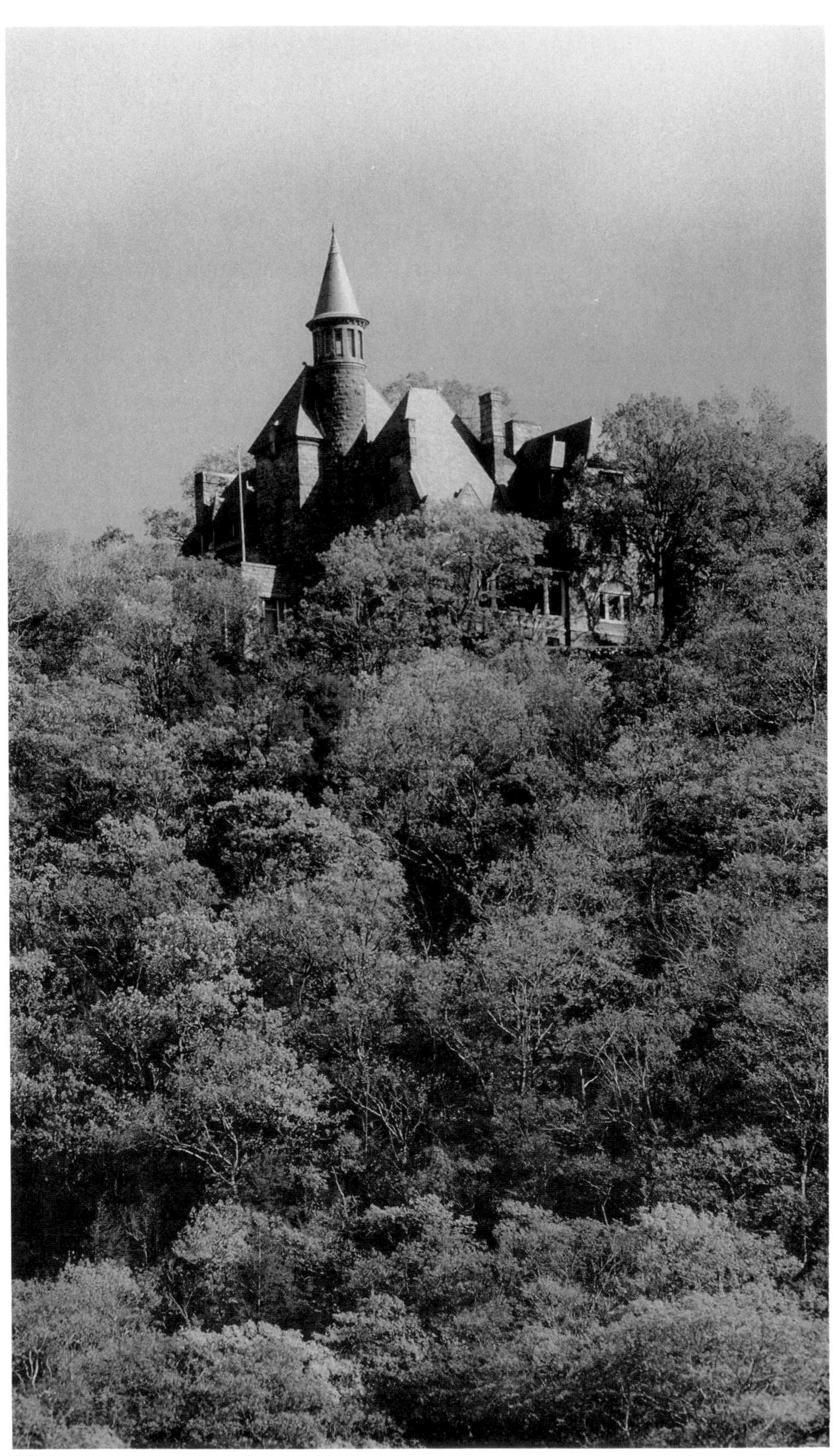

Castle Rock built by William Henry Osborn. Copyright © Charles H. Porter, 1990.

8. Millionaire's Row

THE YEARS following the Civil War have been called the Bare Knuckle Era—the roughest age in the history of American capitalism. This was the age of the "robber barons" whose object of desire and source of wealth was the railroad. In those days, to be a railroad president was to be king, having more power in state politics than governors and more wealth than anyone could have imagined. This powerful engine of capitalism was fueled in wartime by the transport of troops and supplies and later sustained by westward migration and the creation of new markets for goods which had previously been too costly to ship. There were literally hundreds of rail companies formed with government subsidies, earning fabulous sums of money for their owners and promoters.

Among the most powerful of the railroad monarchs were the handful of men who established summer homes in the Highlands. Their vision of country life was as great as their thirst for monopoly. World travelers, they viewed the Hudson as their Rhine, and they built their castles on its shores. Choosing the choicest waterfront properties and the most superb views, they lined the river bluffs with their elegant estates, transforming the Highlands from a resort and health spa to a fashionable suburban enclave. People would call it Millionaire's Row.

The settlement of the Highlands by these families, which had begun slowly in the 1840s grew rapidly in the period from 1860 to 1880. The area would remain under their influence until the 1920s when the railroad industry began its slow decline. By then, several of the wealthy Highlands families had moved on to newly fashionable spots such as Newport and the Adirondacks. However, in the years they made the Highlands their home, they would not only participate in the total transformation of the American economy but also leave a lasting imprint on the region's future and development.

In their professional lives the railroad magnates were consumers of resources. The companies under their control developed the prairie, mined ores, prospected for crude oil and minerals, milled steel, and laid track by the thousands of miles. At their leisure in the Highlands, these captains of industry communed with nature, joining their workmen in tilling the soil, collecting specimens of birds, fish, and plant life, and molding the landscape of their country estates into fine examples of the natural style espoused by Downing. Although Downing had died decades earlier, they clearly found continued inspiration in his idea that creating natural beauty in one's home environs expressed culture and good taste. In their life styles and in their selection of designs and locations for their homes, they sought to demonstrate their bond with nature. From turreted flights of architectural fantasy to understated clapboards, their houses and grounds carried out the Picturesque ideal, making a statement of their social position and of their love for the land. To leave no question, they gave their estates names evoking romantic images of Indians, scenery and wealth—"Oulagisket, "Cragston," "Castle Rock."

The leaders of the new railroad industry living in Cold Spring or Nelsonville included Guy Heustis, controller of the Delaware and Lackawanna Western Railroad; James Heustis, president of the Boston and Maine; and Frederick P. James, who dealt in western railroads. Garrison-on-Hudson, the more coveted address, attracted William H. Osborn, president of the Illinois Central; Samuel Sloan, president of the Hudson River Railroad and later of the Delaware and Lackawanna; and John M. Toucey, general manager of the New York Central. E.H. Harriman, President of the Union Pacific, did not live near the river but maintained a sizeable estate, "Arden" in the nearby western Highlands.

The Highlands also had their statesmen, financiers, and reknowned professionals. Hamilton Fish, who lived in Garrison, was Governor of New York State from 1848 to 1851, when he was elected to the U.S. Senate. He later served as secretary of state in the Grant administration. His son Stuyvesant Fish, who built a house nearby, succeeded William Osborn as president of the Illinois Central. Down the road lived Edwards Pierrepont, minister to the Court of St. James. John Bigelow, an eminent statesman of his time who campaigned against slavery, lived across the river near West Point. Among his neighbors was J. P. Morgan, who by 1910 was the undisputed financial power in the country, controlling most of the nation's railroad lines as well as about fifty major corporations and conglomerates. James Stillman, president of National City Bank and Morgan's business rival, owned land on Storm King Mountain in Cornwall. Richard Upjohn, noted architect of American gothic style churches (including Trinity Church in New York), purchased Mandeville, the colonial house in Garrison which had quartered American troops during the Revolution.

There were a number of reasons why the Highlands were settled by this group of people. Most important was the extension of a railroad line to Garrison in 1848, which made it possible to commute to Wall Street. During the sweltering summer months, a businessman could live in the Highlands with his family, arrive in the city in time for business hours and be back home in the country for the evening. The Highlands had other advantages. Land was cheap, and the mountain air was healthy, offering a refuge from the disease-ridden city. Any one who could afford to get out of New York wanted to. Finally, most of the land suitable for estates had been settled to the north and south by the Hudson Valley's old wealth, the patent holders who were granted their land by the King of England. The infertile lands of the Highlands had no real appeal for the wealthy until sublime scenery became popular, and then, suddenly, an estate in the Highlands was quite fashionable. The Highlands became a private preserve.

Most of the families who moved to the Highlands in the post-Civil War years were related by business, marriage, or both. Morgan's first wife, Amelia

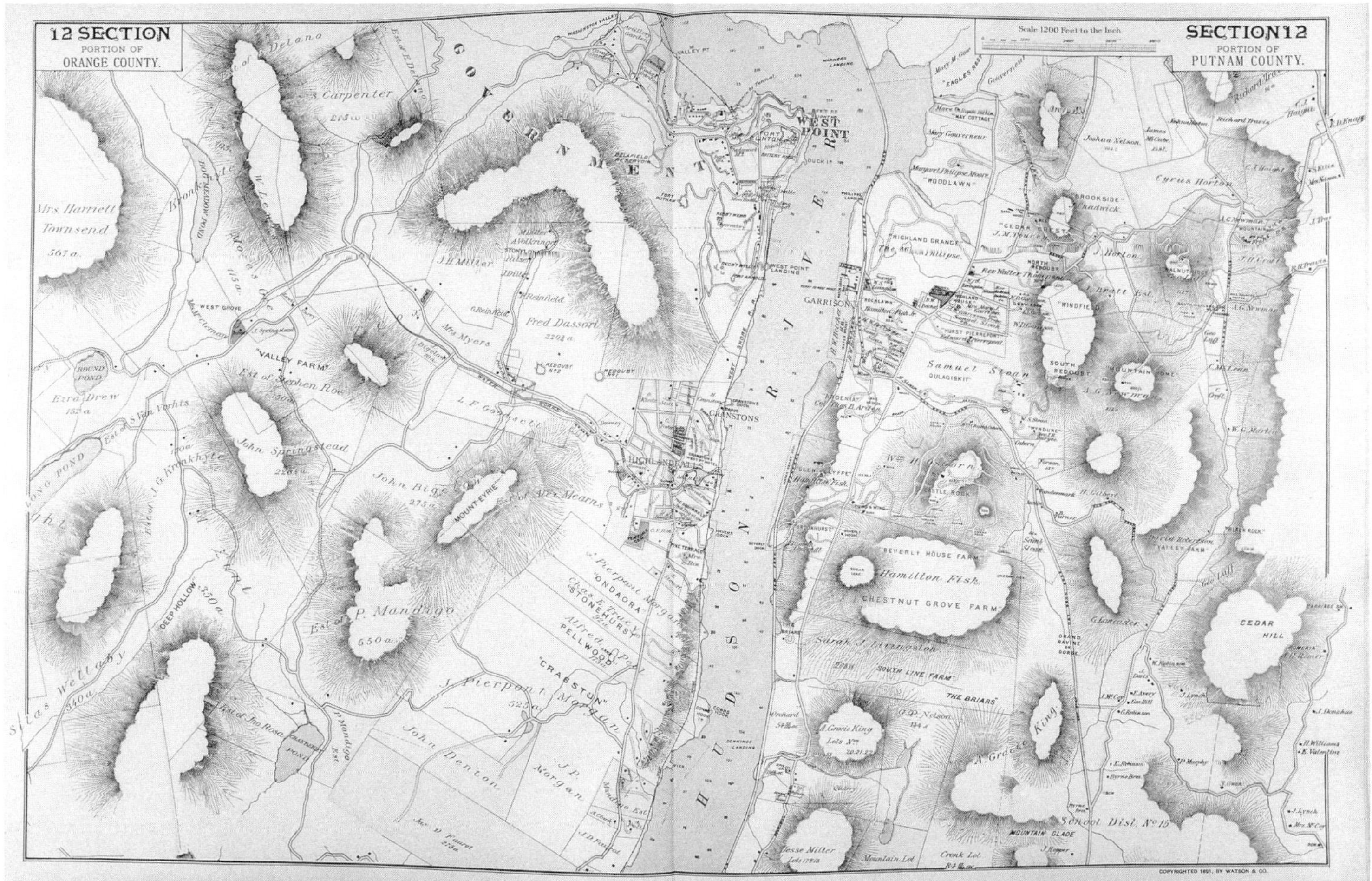

1891 map of the Hudson Highlands. Showing property ownership in Highland Falls and Fort Montgomery, Orange County, from the Beers *Atlas of the Hudson River.*

Sturges, who died of tuberculosis a year after their marriage, was the sister of Osborn's wife, Virginia Sturges. Morgan, who had visited the Osborns in the Highlands, later moved there and bought "Cragston" in Highland Falls with his second wife, Frances Tracy, whose parents, brother and sisters also lived on west shore estates. Charles Edward Tracy, her brother, married John Bigelow's daughter Jenny and lived at "Stonihurst" not far from Bigelow's home, "The Squirrels." Her father, Charles Tracy, who also had a summer home in the Highlands, was a lawyer who handled the Morgan business account. Her sister, Mary, married a Pell. The Pell family had three summer homes in the Highlands adjacent to Fort Montgomery.

Of the nine estates which extended from West Point south to Fort Montgomery, five belonged to Tracys and their relatives. The rest belonged to friends and business associates, with the exception of "Pine Terrace," the estate of General Roe, son of the proprietor of Roe's hotel. Roe was somewhat excluded from the social scene—no one knew where he got his money. The Morgans remained good friends with the Osborns across the river in Garrison, and their children used to try to fly homing pigeons across the river with messages to each other.

The Osborns were also connected to the Fishes and the Harrimans through their roles in the management of the Illinois Central Railroad. Their business lives and the style in which they conducted their affairs were as tumultuous as the bold and rugged scenery they admired.

The building of the railroads was characterized by periods of speculative boom offset by panics and periods of depression. Railroad finances affected the economy more than national finances, and changing fortunes of the railroads set off the Panics of 1857, 1873, 1893, and 1907. The practices of the railroads were cynical and often unscrupulous. The kings of railroading were those who emerged stronger after each panic, many of which they instigated. Retaining power and control often required access to huge personal or corporate financial reserves in order to maintain the credit of the railroad line. In the competitive jungle of railroading, companies slashed their rates until one of them collapsed. The railroads were frequently overcapitalized and underutilized. Separate lines were built serving the same territory, and the railroad barons resorted to ruinous competition to capture the market. The survivor would buy out the competition, usually at discount rates, emerging stronger with each conquest.

To the victor went the spoils. After the Panic of 1893, at least 156 railroads collapsed including some of the greatest in the country (the Erie, the Union Pacific and the Northern Pacific) but from the broken pieces, a few men assembled unprecedented power. By 1906, two-thirds of American railroads had been consolidated into seven great and several minor systems, each controlled by either J. P. Morgan or E. H. Harriman and his banking associates in Kuhn, Loeb & Co.

The Illinois Central was one such line, and its history sheds some light on the tenor of the times and the personalities of the Highland residents who ran it.

The Illinois Central, incorporated in 1850, was near bankruptcy in 1855, when William H. Osborn, son-in-law of one of the incorporators, took over the presidency. Allegations of fraud, among other things, had prevented the completion of the track. Osborn reorganized the company and established it on a firm financial footing, overseeing the completion of the track a year later. Under his leadership, it established new markets for livestock and grain, sold by prairie farmers who had bought land from the railroad as part of a federal land grant policy. By the time of the Civil War, the Illinois Central had expanded into the south and was perfectly positioned to provide for the transport of troops, war materials, and supplies. Osborn retired as president in 1865 but remained active as a director until 1876. He was known as an adventurous and independent man who hated sham and who involved himself in philanthropic activities throughout his railroad career. Osborn and his wife were active in promoting the welfare of railroad employees.

Ten years later, another Highlands resident would take over the line. Stuyvesant Fish had joined the Illinois Central company in 1871. Starting as a clerk, he "graduated" the following year to the office of Secretary to the President. Soon after, he left the Illinois Central to try his hand at banking, but returned to the road in 1877 as a director of the corporation, becoming president in 1887.

The railroad then made most of its profit in the transportation of timber, farm products, and mined minerals. Fish extended the line to create new traffic, building an independent, self-sustained system consisting of main-line tributaries through the Mississippi Valley from the Great Lakes to the Gulf. Fish also established an employee stock-ownership plan, which contributed to the company's success. In his nineteen years as president, he increased the operating mileage by 175 percent, raised gross receipts by 365 percent, and increased dividends by 227 percent.

In 1906, the success of the Illinois Central made it a coveted company for takeover. At the same time, Fish, through an uncompromising stand he took as a member of a committee investigating charges of corruption against officials of the Mutual Life Insurance Company, had antagonized powerful men in the financial community. Fish has been described as a man whose "good-nature and air of *camaraderie* . . . may have led some to underestimate his moral sturdiness, but when the time of crisis came, . . . even the sacrifice of valued friendships could not cause him to waver. Still less could the lure of wealth affect his judgment."[1] When Fish's recommendation for a thorough "house-cleaning" was ignored by the other members of the committee, Fish resigned from the Mutual Life board and went to the press with the facts. Wall Street responded with a threat to remove Fish from the presidency of the Illinois Central. Within eight months he was deposed by vote of the board of directors, and replaced by E. H. Harriman, a director of the Illinois Central who owed his seat on the board to Fish.

If Fish was a man who was loved by "hosts of friends," Edward H. Harriman had no such following. Clearly, he was not a man to tangle with: he was described as "cold and ruthless" in his business relations, "sparing neither friend nor foe if they blocked his plans."[2]

In 1906, when he toppled Stuyvesant Fish from the leadership of the Illinois Central, Harriman controlled more miles of railroad than anyone but J. P. Morgan. At the peak of his power, he personally managed 25,000 miles of line while another 10,000 to 15,000 were under his control, among them the Illinois Central.

Life in the Highlands was a pleasant contrast to the cutthroat dealings of Wall Street. At their country homes Osborn, Morgan, Fish, their relatives and their colleagues delighted in the outdoor life as did Harriman, who had a country estate in the nearby western Highlands. Although much time was spent in dining, partying, and paying calls, the primary activity was enjoyment of the land. Many stayed in the Highlands as long as six months of the year, returning to New York City only when the weather turned cold.

While they were there, they spent hours reworking the landscape and sharing its bounty with guests and neighbors. Susan Hale, who visited the Osborns, wrote in a letter to her sister Lucretia that "The minute we had finished our 1 o'clock dinner, Mr. Osborn took me in the buckboard to see everything. We didn't get back until 6! It is an immense place; in fact, he owns the whole mountain, and he is busy opening vistas and making roads. Of course, the view up and down is perfect."[3]

In the decades following the Civil War, every gentleman had a farm, and every gentleman who had a country seat in the Highlands farmed the land. Since the land in the Highlands was not suitable for farming, the Highlands' new gentry literally created new soil, moving mountains of earth and hauling in manure by the boat load. An 1878 entry in statesman John Bigelow's "Inventory of Homestead at Highland Falls in Orange County" included payments for two boatloads of manure brought up from New York City in a sloop owned by Captain Hiram Meeks. Each boat carried about fifty wagon loads of manure, for which Bigelow paid about sixty dollars.

Bigelow's son Poultney writes about "The Squirrels" in his book, *Seventy Summers*. There, his father was fond of pitching in with the farm laborers. "He entered enthusiastically into the joys of amateur farming—blooded stock, the building of stone walls, erecting new buildings, draining swamps, planting orchards. . . . My father delighted in swinging an axe and helping with a crowbar when rocks were being removed and a wall built."[4] Bigelow came from farming stock in Malden-on-Hudson about fifty miles north and never lost his interest in agriculture. As a statesman, he traveled a great deal, and in his garden at the "Squirrels" is a horse chestnut tree planted from the seed of a tree growing in Voltaire's garden in France. Bigelow, an admirer of the French philosopher, picked up the seed when traveling in Europe and planted it at "The Squirrels," where it is still growing.

Maintaining farms and gardens required a large staff. Lorenz Graber, who owned the florist shop in Highland Falls, remembered that his father had been

"The Squirrels". Copyright © 1990 by Robert Beckhard.

hired to manage the grounds of General Roe's estate. He supervised five other people.

J. P. Morgan was also an active gardener and farmer. He turned the thin, gravel-strewn soil of "Cragston," near Bear Mountain, into a lush estate complete with fruit trees, vineyards, flower gardens, and a dairy. This was no small task. Scores of laborers carted out rocks and brought in humus and topsoil until his 675-acre farm was one of the fattest and most splendid. Morgan delighted in taking fruit, milk, and other fresh products down to the city for his friends. On his 300-foot yacht, the *Corsair*, Morgan held extravagant parties at which he treated his city guests to the fine produce of his country estate—fresh cream, butter, pears, and apples—products that were very hard to come by in the city, where standards of quality had sunk to the lowest levels. (Milk, for example, was watered down and produce was usually rotten—all the more reason to take the family to the country).

The gentry were fascinated with the natural history of the area and helped to collect specimens for naturalists such as John Burroughs, who lived for a time in Highland Falls, and Edgar Mearns, whose monographs on birds and fish of the Hudson Highlands would be published by the Essex Institute in the 1870s and 1880s. Mearns notes that Osborn's son William provided him with an American Goshawk "procured at Garrisons," and Miss Anna B. Warner obtained another on Constitution Island. He credits John Burroughs as the source for observations

"Stonihurst." Copyright © 1990 by Robert Beckhard.

of the blue-winged yellow warbler. In attempting to document the speed at which an old squaw duck was flying, Mearns observed that the bird flew at exactly the same rate as a passing train. John Toucey, superintendant of the railroad, was pleased to provide the times and distances to Mearns who then figured that the duck was flying at the rate of about 1 mile per 1.43 minutes.

Reverence for nature did not preclude killing wildlife; procuring specimens generally involved shooting them, and Mearns describes the sighting of a laughing gull near Constitution Island that "was so gentle that a boatman attempted to strike it with an oar." The gull wisely flew away. The wildlife of the Highlands also added romantic effect, and some species which are now rare were then abundant. Mearns offers this description of the bald eagle, which he found to be a permanent breeding resident:

> The White-headed or Bald Eagle constitutes a marked and romantic feature of the superb scenery of this part of the Hudson, lending another charm to a scene already grand and impressive, but rendered sublime and awe-inspiring by the presence of this noble bird, seen perched upon some blasted tree above the massive cliffs, or soaring in higher atmospheric regions, far above the reach of the coming tempest, while its scream falls faintly upon the ear, answering the loud, quavering cry of its nearer mate.
>
> In winter, when the river is frozen, the Eagles are seen soaring above the mountainside, searching for the scanty prey upon which they are obliged to

> subsist when fish, their favorite food, is unattainable; but later, when the ice is in motion in the Hudson, carried swiftly by the current, numbers of them may be seen sitting in pairs upon trees low down by the river's edge, watching for their finny prey, or else floating upon the ice in the stream, in company with Crows and Gulls. In summer, their favorite perch is upon some withered tree on the mountain's side, from which, at intervals, they descend to the river, or some secluded lake, to seek their food. When the ice first breaks up in the Hudson, the Eagles are sometimes extremely abundant. At that season I have counted more than twenty-five that were in view at once.[5]

Most other entries in Mearns' Hudson Highlands monograph are the matter-of-fact observations of a naturalist. However, this passage, published in 1880, shows that the concepts of the sublime and picturesque which took hold in the early 1800s still prevailed in the 1880s. This was part of what made the region so appealing as a place to build a country seat.

Family life at the great estates has been beautifully recorded by Grace Tracy Cook who describes the Bigelow and Morgan households in her memoirs. A granddaughter of Bigelow and a niece of Morgan, she spent her childhood at "Stonihurst" in Highland Falls at the turn of the century, when activity at these estates was greatest. She recalls that "Uncle Pierpont" Morgan was given to a royal display of wealth at "Cragston." His servants were dressed in striped waistcoats and gorgeous livery.

She wrote:

> Cragston was kept open for weekends a good many years and there was a great deal always going on there when the house was open. They lived in the most lordly English style. In the spring and summer Uncle Pierpont came up for the

"Cragston" interior.

> weekend on the Corsair, the palatial yacht which we could see at anchor from Stonihurst. He used to drive a coach and four horses to church and often picked us up—young and old—and the young were hoisted onto the top while he drove, and the awe-inspiring coachmen and footmen in livery and top boots sat behind. Driving through the village thus is an experience I have not had since—nor do I think I want to. . . . The meals at Cragston were positively gargantuan and Uncle Pierpont's rule was: No breakfast in bed for anyone unless ill. Sunday morning everyone trooped down for a seven-course breakfast at ten. After a gigantic lunch served for guests (which he never ate) there would be an English tea spread. That was followed by Sunday evening supper, equally gigantic to which family and neighbors were invited. All this was followed by Hymn Singing which he chose. Any hymn was acceptable except 'Nearer My God To Thee'[6]

By contrast to the opulence at "Cragston," life at John Bigelow's estate, "The Squirrels," was more down to earth. Bigelow, an eminent political and literary figure, was not in Morgan's financial league. Grace Tracy Cook remembered that the Bigelow grandchildren ran barefoot in homemade brown linen dresses and were quite awed by the pomp and ceremony at Uncle Pierpont's. Bigelow believed strongly in family life, health, and moderation. He was a proponent of homeopathic healing and did not believe in conventional medications such as quinine, aspirin, or laxatives. He treated his family, his servants, and even his horses with homeopathic methods. "*No one* played cards, ever had an operation, got drunk, nor cared for anyone's opinion but his or her own," wrote his granddaughter.[7] Politically, he was antislavery and a passionate advocate of such issues as free trade and Panamanian independence. At "Stonihurst," the Bigelow family sewed a flag for the new republic of Panama.

Bigelow had moved to the Highlands in 1856 out of concern for the health of his son Poultney, an asthmatic baby struggling for life. It must have helped; Poultney lived to ninety-eight. When he first purchased "The Squirrels," a twelve-acre property, Bigelow was owner and editor of the *Evening Post* along with William Cullen Bryant. Soon after, he was appointed consul general at Paris. In 1867, when he returned to the United States, he campaigned for New York Governor Tilden, and was later New York's Secretary of State. An editor, author and diplomat, Bigelow was a member of the commission which broke up the New York canal ring and an expert on World War I diplomacy. In his later years, he was a primary organizer of the New York Public Library and first president of its Board of Trustees. Compared to the railroad barons and other capitalists who lived on the west bank, he was not a wealthy man. With strong political and literary leanings, he shunned the financial wheelings and dealings of his friends and relatives.

Grace Tracy Cook writes of her grandfather's home:

> the Squirrels still remains to me the past's most enduring reminder—Grandpa in his light homespun and a light blue necktie, slowly walking up and down the garden path to the summer house at the end, that hung practically over the cliffs above the river. . . . The occasional sweltering summer nights. . . . The

John Bigelow with one of his granddaughters. Courtesy of Constantine Sidamon-Eristoff.

> balmy days of September, when the sky and the river seemed to become a sea of blue, in which the mountains floated, the day boat and the night boat with its search light. . . .[8]

Each of the nine estates in Highland Falls and Fort Montgomery had extensive gardens, and it was customary to allow "The Village" to walk through them all on Sunday afternoons. It was possible to walk from one to the next through all nine. Young officers paid their calls on horseback and remained for supper. West Point was an important part of the social scene for those who lived in the Highlands and was another reason why the captains of industry found the region attractive. One could enjoy the outdoors and at the same time socialize with the elite group at the academy. Enthusiasm for boating also added to the pleasure of living in the Highlands. The steamboat made it possible to entertain summer guests and visitors, and the estates also had their own docks for yachting. Again, Grace Cook offers us a clear picture:

> Only my generation can remember the simplicity and innocence of West Point when I was seventeen. There were three hops a week, all summer, which began at eight and ended at ten, with white gloves for all, and lemonade. The cadets were never allowed off the post at any time, and the post was then half the area that it is now. Hops were card dances and on the whole quite solemn and it was only on these occasions or Sunday afternoons before Dress Parade that a cadet could see anything of the opposite sex. They were never out of dress uniforms except for drills and recitations. It all had an aura of discipline both social and intellectual and was the only place in the North where tradition was not only preserved but lived by. The army was still more or less what it had been in 1804. The democratization of it began with the mechanization of the cavalry after the 1st World War. The old charm has completely gone, in my opinion.
>
> . . . West Point was at that time considered an ideal summer residence for young ladies. It was easy for chaperons, and the steamer Mary Powell, alluded to by every writer about the Hudson, furnished a cool and comfortable means of transportation from New York. I remember the Mary Powell well. Horses and carriages could be put on board her, and it was an easy way to travel as far as Albany and the Catskills and even the Adirondacks, although they were not discovered until later. There were weekly dances at Cranston's; the proprietor probably had fewer headaches than the management of a summer hotel has today. Most of his food supply was local. Beginning after the Civil War, Highland Falls came to be considered fashionable, if you can believe it, and little by little summer houses were built near it—or existing farm houses bought and adapted by people who came for the summer. In the 1840s country houses were built by three members of the Pell family. They were the first to put up what was then considered 'handsome' houses with docks fronting on the River . . .[9]

Although Morgan and Bigelow had townhouses in New York (Morgan's at 219 Madison Avenue was unmarked; everyone was supposed to know he lived there), they also carried on business from the country. Both men entertained prominent guests at their country homes, among them Oscar Wilde, who visited Bigelow at "The Squirrels" on his first lecture trip to America. The State Democratic Committee came to Highland Falls to ask Bigelow to run for Congress, which he declined.

During the time Morgan summered in the Highlands, he was approaching the peak of his power. Edward Mitchell, editor of the *New York Sun* wrote of him "The lesser monarchs of finance, of insurance, of transportation, of individual enterprise, each in his domain as haughty as Lucifer, were glad to stand in the corridor waiting their turns like applicants for minor clerkships in the anteroom of an important official, while he sat at his desk in his library within."[10]

Railroads were not his only financial interest. Out of the chaos of panics and mergers, J. P. Morgan created his own brand of order and control, particularly in the transportation and banking industries. He connected the previously unconnected railroad and steamboat lines and created the first regional, then national, train systems. Among Morgan's creations was the U.S. Steel Corporation. He financed the Edison Electric Luminating Company, and with the help of his banker, James Stillman, he forged the Consolidated Gas Company out of many smaller utilities. After his death, in 1936, these companies merged to become the Consolidated Edison Corporation. Among the other great businesses financed by Morgan were the International Harvester Company, the National Tube Company and the American Bridge Company. He dominated or controlled General Electric, American Telephone and Telegraph, and scores of other companies.

While life at the country homes of Morgan and Bigelow clearly revolved around family and fun, a different drama was acted out at James Stillman's estate in Cornwall. Stillman moved to the Highlands in 1885. Soon to become president of National City Bank (1891), he was a business ally of E. H. Harriman and William Rockefeller. These men were adversaries of J. P. Morgan until 1907, when rivalries were set aside in favor of the "community of interest." Snubbed by the Tuxedo Park set, who regarded him as *nouveau riche*, Stillman decided to build his own, better compound in Cornwall, where he had attended boarding school as a boy. He purchased a large tract on Storm King Mountain which he hoped to transform into an artists' colony and country hideaway for his family and friends.

He named his country home "The Ridge." It was a "low, rambling comfortable house,"[11] but it was ruled by the master as though it were a cell block. Stillman insisted on printed menus for his meals. He rated each course of his dinner by assigning a percentage which he communicated to the chef. Like J. P. Morgan, Stillman commuted from Cornwall to New York by yacht. For his departure, he required his family and servants to line up in the main hall at eight in the morning while he marched past. Finally climbing into the carriage, he would be driven to the boat landing. After Stillman's yacht, the *Wanderer*, had left the dock, Mrs. Stillman, who found life with her husband constraining, would reportedly rush to the home of a neighbor and madly play the piano until she got him out of her system. In 1894, Sarah Stillman left her husband and took up residence in Europe. James Stillman told her to never speak to him or the children again and ordered the children to never mention her name.

James Stillman began accumulating his wealth in 1875, when he was twenty-five. His father died in that year, and he became the sole trustee of a million-dollar family inheritance. Within five years, Stillman had made himself a rich man as a cotton merchant. He later invested in railroads and oil. In 1891, he became president of National City Bank, which under his leadership became the top-ranking bank on Wall Street. It played a major role in the sales of stocks designed to secure monopolies for the copper industry and the gas utilities. His estate was valued at $50 million at the time of his death in 1918. A reticent man, he hated publicity and was described as cold and unemotional. At "The Ridge," his hobby was raising beef cattle, and he spent hours photographing them. During the final years of his life, he spent generously from his private fortune to support France in World War I.

Though the landed gentry of the Highlands shared a common interest in the river scenery, the designs of their houses were extraordinarily varied. Some of them, like Morgan, moved into existing farmhouses and converted them to their needs. Others built new villas, usually following the guidelines promoted by Downing, who felt that many different styles could be accommodated in America as long as they were designed with the picturesque or beautiful elements which would complement the landscape. William Henry Osborn, who first became acquainted with the Highlands as a summer boarder at the West Point hotel, where he vacationed with his friend J. P. Morgan, took this notion to the extreme. "Castle Rock" was not just Osborn's home, it was his fantasy. Built on a mountaintop in Garrison, it is tall, turreted and whimsical—a stunning example of picturesque architecture.

"Castle Rock," at an elevation of about eight hundred feet, commands a spectacular view of West Point and the twin sentinels of the Hudson River gorge—Storm King and Breakneck. It looks down on a sweeping panorama of the river and its Revolutionary War sites, and on a clear day, the Catskills and the Shawangunks are visible. Built on the crest of a ridge, it is silhouetted against the sky much like a romantic castle on the Rhine, epitomizing the picturesque ideal so popular among Americans at that time. It is two and a half stories high, constructed of rough-cut granite masonry. Asymmetrical in shape, Castle Rock is dominated by a round tower in the center and capped by a red slate roof. Porches and galleries on the west and south sides frame the views of the valley below.

When Osborn built "Castle Rock," some thought he was lording it over the Fish family, who lived at "Glenclyffe," which was situated on lower ground in Garrison. While that may be the case, the location and design were clearly influenced by Hudson River School painter Frederick Church, a childhood friend of Osborn's. The two had grown up together in Connecticut and remained close throughout their lives. In the 1960s General Frederick Osborn, William Henry's grandson, was instrumental in securing the preservation of "Olana," Church's home and studio near Hudson, New York, as a state park and historic site.

It was Church who suggested the location for the castle which Osborn built in 1881, and he took a strong interest in its design. Church had definite ideas about architecture and had personally designed "Olana" in the style of a moorish castle with features planned for artistic effect. At "Olana," colors change from room to room and every window offers a striking vista. In 1868, Church wrote to Osborn from Rome: "Don't settle any plans about building a house until I return—I am conceited enough to wish to thrust my finger into that pie and offer my opinions on domestic architecture." Less than a week later, he wrote again, saying "I should be almost as much interested in your house as my own and as I have scraped together some good ideas, as I think—I want you to have the benefit of them." In the archives at "Olana" is a pencil sketch done by Church of the proposed Osborn castle.[12]

"Castle Rock" was completed in 1881. Architect Jarvis Morgan Slade was hired to render the design. In 1882, upon Osborn's retirement, the Osborn family took up permanent residence. The building was later expanded by Osborn's son, Henry Fairfield, who added a north wing and library in 1906. When completed, "Castle Rock" boasted thirty-four rooms and four interior chimneys. Guests were invited for stays of a week at a time, and no sooner would one set of guests leave than another arrived.

Susan Hale, an acquaintance of both the Churches and the Osborns, visited the castle three years after it was built. Afraid of heights, she was not enthusiastic about staying there. She descibed her visit at length in another of her letters to her sister Lucretia:

> "Castle Rock" on the Hudson
> Tuesday, July 29th '84.
>
> Dear Luc,
>
> . . . Emulating Mr. Church's palace, Mr. Osborn (who is a "Railway King") has built a castle. I think it is a dreadful place, for I could scarcely bear the Churches being up so high. . . . This is of course very fine, but 'tis wearisome to get up here by the long winding road, ... The Hudson, to be sure, is at our

"Glencliffe," the riverfront estate of Governor Hamilton Fish, from an engraving in Pelletreau's *History of Putnam County.* It was subsequently the home of his son Stuyvesant Fish. The structure has since been remodeled for use as a convent.

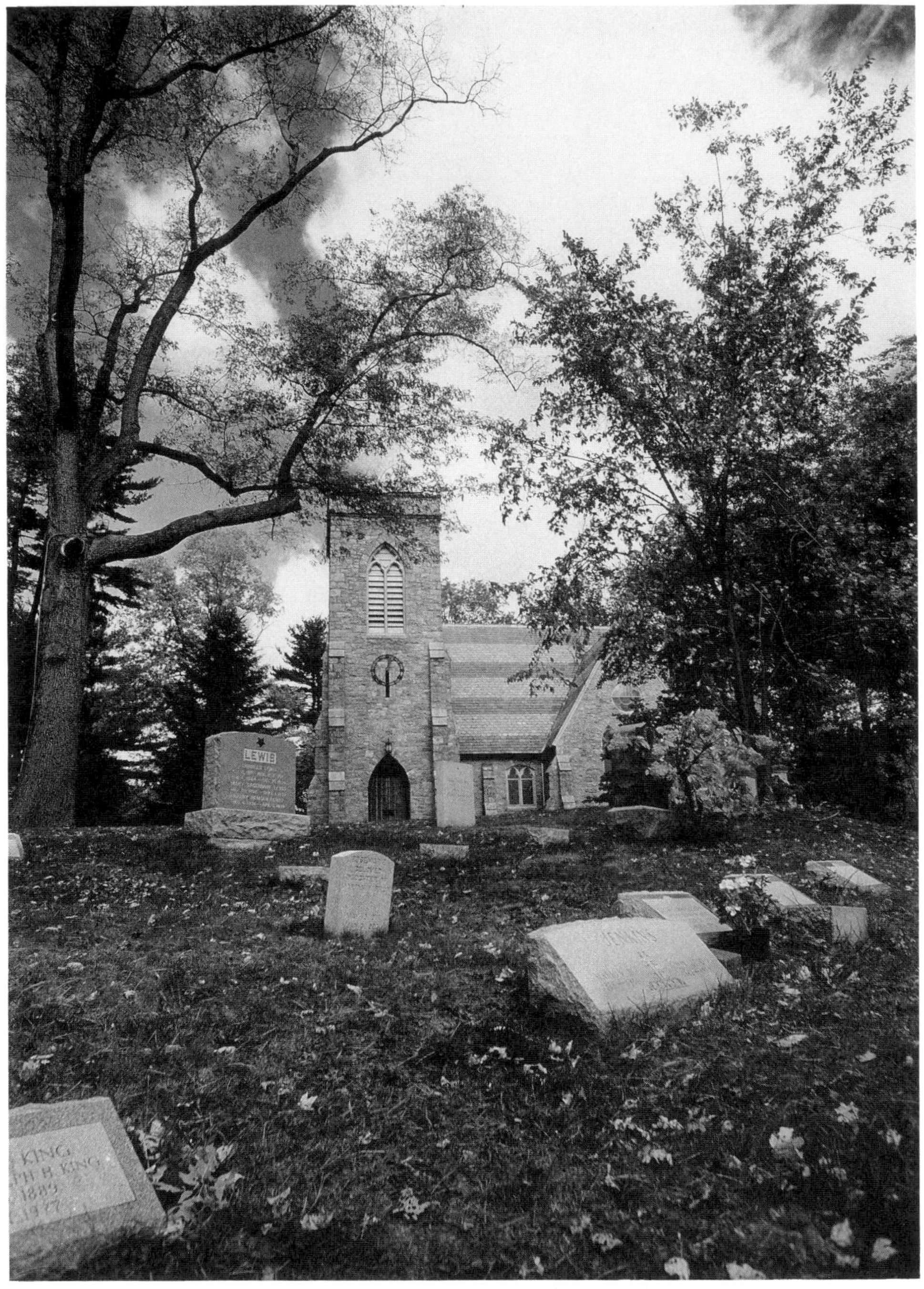

St. Philip's Church in the Highlands where the captains of industry worshipped. It was designed by noted architect Richard Upjohn. William Henry Osborn, Samuel Sloan, and Stuyvesant Fish are among those buried here. Copyright © 1990 by Robert Beckhard.

> feet, and West Point just opposite affords a human interest not got at Olana; but they say here that often thick clouds below this house hide the whole landscape and the sun is shining here while they hear the . . . lamentation of cars and cows . . . from the world below where it is raining. I fervently long to be on terra firma, and have a kind of permanent vertigo.[13]

Some of the Highlands estates were built new, while others such as "Cragston," "The Squirrels" and "Glenclyffe" were houses originally constructed in the

1840s and 50s and remodeled by their new owners with greater or lesser success. Invariably, however, the views were spectacular. "The house at Glenclyffe was never beautiful," wrote Hamilton Fish's grandson.

> like "Topsy" it just grew. After each "reconstruction" (addition) it became more hideous. Coming up the drive which was shady and pretty, the great pile of red brick slapped you in the face, right in the glare of the sun, with an almost treeless lawn in front of it.
>
> The grounds were lovely—the glen where the mint . . . for making juleps grew, "Picnic Rock" . . .—the walks—garden and greenhouses.
>
> I was once forced to go and see the Taj Mahal by an enthusiastic friend at sunrise, noon and sunset. I would never do that again, but I would like to see the view from the south piazza of "Glenclyffe" as it used to be before railroads and bridges marred the landscape—it was gorgeous.[14]

Among those who built new houses in the Highlands, many of the captains of industry and finance employed architects to design their estates, choosing the best known profesionals of their day: Calvert Vaux, Frederick Withers, A. J.

"Tioranda" built in 1859 for shipping magnate Joseph Howland was designed by architect Richard Morris Hunt. Copyright © 1990 by Robert Beckhard.

West Point from Garrison. Copyright © 1990 by Robert Beckhard.

Davis, Richard Morris Hunt, and Richard Upjohn. Others selected their designs from the numerous pattern books then available, perhaps sketching a tower here an arch there or an extra window where the view was finest, and then turning the sketch plan over to a carpenter. The styles, representing an ecclecticism seldom found elsewhere, are unified in their harmonious adoption of a naturalistic style and expressed the views of the owners in relation to the land.

This philosophy of architecture and landscape was shared by friends and associates at West Point. In 1902, when the grounds of the academy were slated for a total overhaul, a design competition was planned. The academy was then celebrating its centennial. Invitations were extended to the leading architectural firms in the country which had expertise in the styles characterizing the academy at that time—English Tudor, Classic, and Eclectic. The competition became known as the battle of the styles. The choice of a style and plan for the expansion of West Point would have a lasting impact on the river because of its size and location.

In 1903, a West Point jury selected the firm of Cram, Goodhue and Ferguson of Boston for their plan in the Gothic style, one which differed from any of those at the academy but which captured the romantic and naturalistic qualities of the site. In their winning proposal, the architects noted that their aim was to preserve the "natural features which give West Point an extreme distinction of landscape, and to "emphasize rather than antagonize the picturesque natural surroundings of rocks, cliffs, mountains and forest and be capable of execution at the smallest cost consistent with the monumental importance of the work."[15]

Construction began in 1903 and was completed in 1914. Among the most notable structures which date from this period are the Administration Building, which with its pointed arches, stone window tracery, ribbed and vaulted ceilings, buttressed walls and Gothic ornamentation, resembles a medieval dungeon; the East Academic Building; the Riding Hall and the North Barracks. In the Riding Hall, the ramparts of stonework rise from the water's edge like the Hudson's Palisades to the south of the Highlands. Nearly all the buildings erected during this period utilized rough-cut stone from local quarries which made the structures harmonize with the prevailing colors of the locality.

The Highlands were particularly appealing as a location for castle building, because they offered outstanding views and a setting much like Germany's Rhine. A castle added to the romantic character of the Highlands. Including Osborn's Castle, there were three of them. In 1903, employing modern construction methods, Evans Dick and his wife, a daughter of Stuyvesant Fish, started construction of a castle built of cast concrete. Located on a mountaintop in Garrison, its design was inspired by Granada's moorish palace, the Alhambra. The plan called for 52 rooms, including a ballroom. During the construction, more than 100 artisans and workmen camped on the mountainside for about eight years. In 1907 a stockmarket crash depleted the Dick family's wealth, and in 1911 work was halted with only $3 million spent of the projected $7 million cost. Building materials, ladders, and workmen's tools were left on the site as if work would resume the next day. However, the castle remained in its incomplete

Architectural detail of Dick's Castle, made of concrete. Copyright © 1990 by Robert Beckhard.

Bannerman's castle, built on Pollepel Island by Francis Bannerman for use as a "baronial" castle and as an arsenal for storing munitions for his New York city surplus weapons business. Now part of Hudson Highlands state park. Copyright © 1990 by Robert Beckhard.

state through the 1930s when Mr. and Mrs. Dick died. (It has since been renovated by an investor and turned into condominiums.)

In 1900, Francis Bannerman, a New York entrepreneur who was one of the world's largest dealers in second-hand military equipment, purchased Pollepel Island and built an "ancestral" Scottish-style castle on it. It was completed in 1908 and used as both a home and an arsenal. Bannerman's company stored powder and ammunition there until 1920, when it exploded amid a barrage of shot and bullets and a dense cloud of smoke.

It was an appreciation for nature and scenery that brought the captains of industry to the Highlands, and it was the railroad that made it possible by building family fortunes and establishing a commuter service. While the railroad giants left a lasting imprint on the landscape of the region, the construction of the Hudson River railroad line had dramatic impacts of its own, filling shallows and wetlands and virtually severing the river's bays, inlets, and tributaries from the main stem. The railroad formed a steel barrier between the land and the river and blocked the sloop trade from its access to markets. Where once the wharf had been the center of activity, it was now the train station, and development began to take on a linear character, straight and unswerving as the rails themselves.

The railroad on the east shore of the Hudson was built in stages between 1847 and 1851, and the west shore line was constructed about thirty years later.

Nationwide, more than 15,000 miles of track had been laid before the east-shore Hudson River line to Albany was completed. On a river of firsts, the railroad was close to last.

When a line between New York and Albany on the shores of the Hudson was first proposed, investors viewed it with caution. Construction of a railroad along the Hudson posed engineering, political and marketing difficulties. The steep mountain slopes of the Highlands had always been a barrier to transportation. A shoreline route would have to carve a foothold at the base of the mountains. There was also the power of the steamboat operators to be considered. They held what was effectively a transportation monopoly on the river, and they had friends in high places, including the state legislature, where any railroad corporation would have to be chartered. Most problematic was the question of whether a railroad could successfully compete for passenger and freight traffic against the luxury steamboats so adored by the public. At the time the railroad was being proposed, there were twenty boats in regular daily operation between New York and Albany. As one historian has described it: "The Hudson had become one of the busiest waterways in the world. There was scarcely a mile of it without great, graceful white steamboats plying their swift and placid way. The river towns were agog with the gossip of the lines. Their

Hudson River Railroad, Anthony's Nose, anonymous woodcut, 1853. Courtesy of the New-York Historical Society, New York City.

taverns and their shops echoed with it. The line runners swarmed the wharves. Captains were men honored and respected. Their passengers were folk to be envied. This was the golden age of steamboating."[16] Steamboats offered everything in the way of food and drink. They were noted for their vast saloons and lavish dining rooms. It would be decades before the railroad car would offer similar comforts.

Though Osborn, Morgan, and other railroad magnates would move to homes along the Hudson after the Hudson River line was built and would profit by the commute to New York City which the line provided, they were not among its backers. Financially, a railroad along the Hudson was a long shot, and the difficulty of raising the necessary three million dollars in capital was sufficient to discourage most from trying. The issue was discussed in 1842 by a group of Poughkeepsie investors. They attempted to obtain a charter from the legislature. However, steamboat interests quashed their efforts. Undaunted, they regrouped in 1845 and hired engineer John B. Jervis to conduct a preliminary survey of a route. They offered him only a small fee for his services, but guaranteed that he and his assistant engineers would be employed if the project should succeed.

Jervis did a market analysis which he presented in a report to the committee on January 20, 1846. In his study, Jervis estimated the number of passengers and the ticket rates which could be charged, while still remaining competitive with the steamboat, Jervis recognized the railroad's advantage in being able to operate year-round, compared to the steamboat which could only run about nine months of the year, when the river was not frozen or choked with ice. He guessed that 250,000 "through passengers" and 250,000 "way passengers" would ride the train each year, if the trip from New York to Albany could be run in five hours at a ticket price of about one cent per mile. This was the highest fare it was thought the railroad could charge and still compete with the steamboat. With additional revenue from freight, Jervis concluded that the road could produce nearly half a million dollars in net revenue per annum. At about seven percent, he deemed it "a fair return for the capital invested." His presentation was received with considerable enthusiasm but failed to produce what Jervis described as the "one element essential to such projects—that is, confidence in its ability to recoup the outlay required. All were ready to speak favorably, but very few ready to take any pecuniary responsibility."[17]

In February, Jervis personally took the proposal to Albany, seeking an act of incorporation. At the Capital he was faced with the opposition of both the steamboat interests and the backers of the New York and Harlem railroad, a potential competitor. After three months of political struggles, the act was passed. In June 1846 a board of directors was formed which included Jervis and Cold Spring's foundry president, Gouverneur Kemble, among others. A prospectus was issued, seeking subscriptions to the stock in the Hudson River Railroad Company. The prospectus projected a return of seven percent on investment and pitched the railroad's superior rate of speed compared to that of the steamboat; its access to manufacturing establishments and agricultural areas for freight service; and the potential for growth of commuter traffic through development of the Hudson Valley:

> The banks of the Hudson afford to an indefinite extent, the most beautiful situations for country residences. The proposed Road will afford to a numerous class of our citizens, who for themselves or families, desire to reside in the country during the summer, the greatest possible facilities to do so; allowing them to reside 20 to 80 miles from the city, and still be in it during the business hours of the day, and return to their families at evening."[18]

Jervis and his backers still had some convincing to do, however. Except for themselves, the stock had no takers. It was not until nearly a year later, on March 1, 1847, after a lengthy publicity campaign, that the three million was raised and work began.

With Jervis as chief engineer, the land was purchased and the track was laid with astonishing speed. Work was commenced in sections, with phase I covering the fifty-three miles from 32nd Street in New York City to Breakneck Mountain. Construction conditions were much more difficult than on inland lines. To create a straight line across the Hudson's bays and inlets, a significant amount of track was laid on rock-fill in the river. This changed the east shore of the Hudson environmentally as well as physically. For much of the distance through the Highlands, where the river was deep, a bench of rock had to be cut from the steep-sloped shore. At Anthony's Nose and Breakneck, the only straight path was by tunnel.

The section of the line under Breakneck was one of the most difficult, because the granite proved particularly resistant. It was harder to drill than any rock Jervis had previously encountered. Workmen drilling all day only cut one to two feet into it. Numerous contractors started the work and were forced to give it up. By the middle of 1848, $6000 worth of work had been done, with $150,000 yet to be done. Soon after a new contract was given to H. D. Ward & Company, which completed the work in seventeen months.

Passenger service was extended north to Poughkeepsie on December 1, 1849, with stops at Garrison and Cold Spring. A passenger ticket from New York to Cold Spring cost seventy-five cents. The price for commuter tickets was set at forty-three cents each, for 120 tickets usable over a six-month period. In 1851 the line to Albany was completed and rates were raised about fifty percent. The number of trains increased from two per day to six.

In the first three months of operation alone, passenger traffic totaled 765,877, far exceeding Jervis' estimates of 500,000 passengers per year. In 1850, flush with this triumph, Jervis wrote that the experiment

> has settled the great question, that a well built railroad can successfully compete with steamers on the very superior navigation of the Hudson River, in the transportation of passengers; and consequently they will be required along all the great channels of steamboat navigation. We shall no longer look to the steamer, as heretofore, as the perfection of traveling; but shall cast about to ascertain what facilities are at command to obtain the superior conveyance . . .

> We look to the southern shore of Lake Erie, as one of the first cases that must receive attention. Something is doing on this route; but it will not suffice until a first class road is extended from Buffalo to the head of Lake Erie, and thence by the best and most direct route to Chicago, St. Louis and Galena.[19]

Jervis boasted too soon, however. The railroad was unable to compete successfully for freight, despite its triumph in attracting passenger traffic. The Hudson River Railroad did not turn a profit for fifteen years. Per mile, it had been the most expensive railroad ever built, each mile costing about $80,000 for its 143-mile distance. Jervis had grossly underestimated the cost of operation, projecting about eighty cents per passenger per mile, when in fact the cost rose to a dollar and fifteen cents. To turn this situation around, in 1855, Sam Sloan, who lived at "Oulagisket" in Garrison, was brought on as President of the road. He immediately began cutting operating costs and slowly his measures began to achieve results. In 1862, the corporation declared its first profitable year, and Sloan's salary, which started at $5000 in 1855 was increased to $10,000.

With the issuance of its first dividend, the Hudson River Railroad was finally declared a success, causing Thurlow Weed to write in the *Albany Journal* in 1863: "We were originally among those who could not believe it would ever be built—who thought it irreverent to attempt to rival God's munificent, glorious highway, the Hudson River. If our files were searched we should be found expressing our opinion that the idea of a 'Railway to the Moon' was scarcely less preposterous than the projected one along the banks of the Hudson River."

During Sloan's nine-year tenure, stock prices in the Hudson River Railroad Company rose from $17.00 per share to $140. The line later merged with the New York Central, and John M. Toucey, also of Garrison, was appointed to manage the combined lines. Along with the Illinois Central, the newly created New York Central and Hudson River Railroad was one of the few railroads to survive the depression of 1893 to 1896, which caused the collapse of 156 rail lines.

In 1892, Sloan had a new railroad station built in Garrison. It was fancier than the old one and more suitable as the station of departure for so many railroad kings. Morgan, Sloan, Osborn, and Toucey were known to enjoy long chats on the commuter train.

Garrison Landing, the site of the ferry crossing between Garrison and West Point, was the first train stop in the Highlands on the Hudson River line. The landing had been a hub of activity even before the railroad was built. Once railroad service came to Garrison, in 1849, Garrison Landing grew substantially to accommodate the new passenger and freight traffic and became a bustling transportation center. Henry White Belcher had started a ferry service in 1821 providing daily connections between Garrison and West Point. After 1849, the ferry linked the river estates, West Point, and the railroad.

While the construction of the Hudson River railroad had a certain romantic appeal, much the same as the introduction of steam navigation forty years

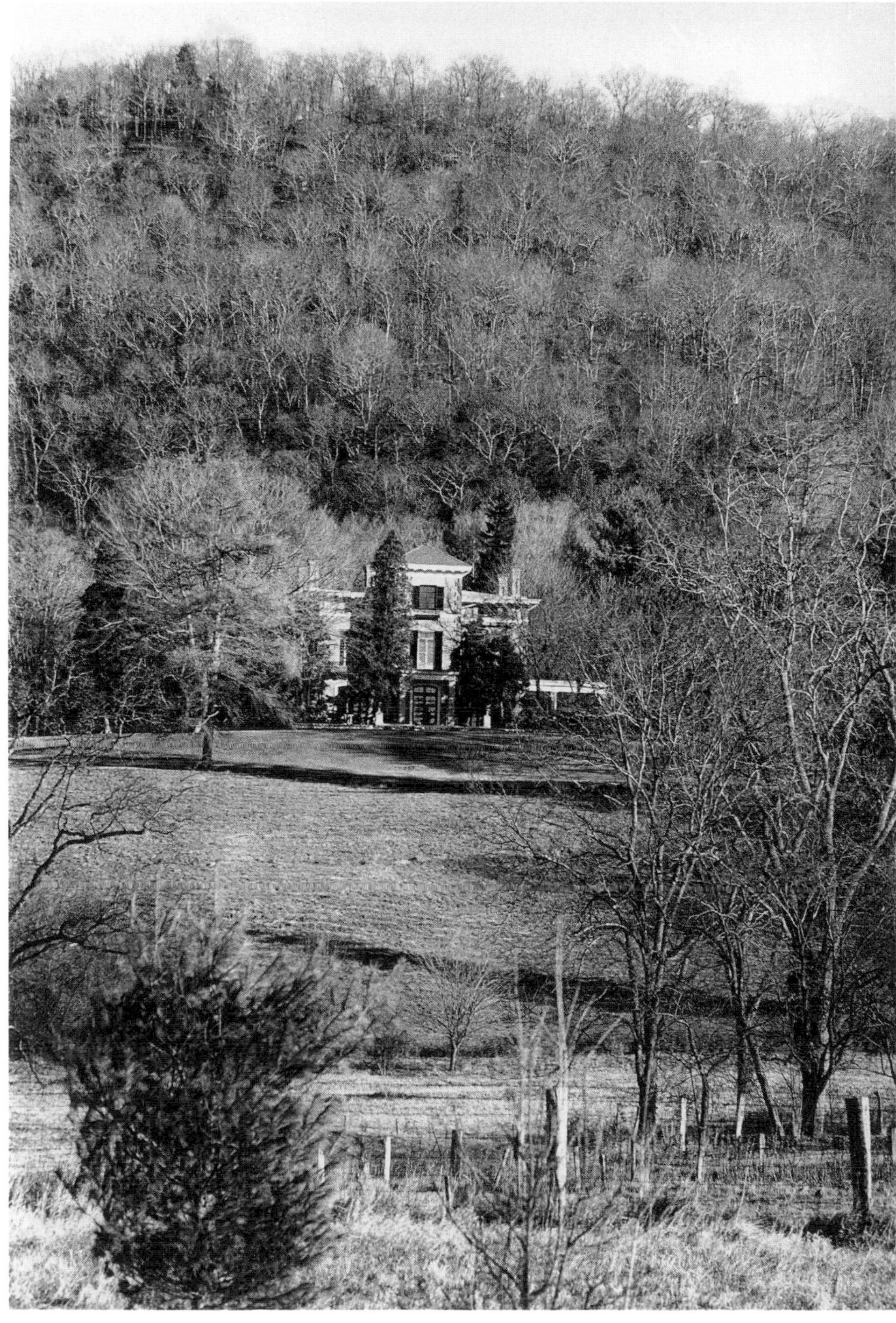

"Oulagisket" the estate of Sam Sloan, President of the Hudson River Railroad. Built in 1864. Copyright © 1990 by Robert Beckhard.

earlier, it produced the first expressions of environmental concern over the loss of scenery and the intrusion of industry. In an article for *Hunt's Merchant's Magazine*, Jervis remarked on the intensity of public reaction which his proposal had inspired: "To adopt a river route, was claimed in the style somewhat of the Spanish Don, to be a *desecration* of the river, marring its beauty, and subverting the purpose of the Creator."[20] Jervis later summarized the debate on this issue in

View of West Point from the Landing at Garrison, anonymous lithograph, 1860. Courtesy of the New-York Historical Society, New York City.

his *Reminiscenses*: "The landed proprietors along the route were in general very hostile to the enterprise, declaring that it would destroy the natural beauty of the country as well as fail in its commercial object. On this point I had claimed that the natural scenery would be improved; the shores washed by the river would be protected by the walls of the railway; and the trees, no longer undermined and thrown down by the surf, would grow more beautiful; and that the railway thus combining works of art with those of nature would improve the scenery. For this I was freely reproached as scarcely less than a barbarian." One of his opponents may have been a man named William J. Blake, who wrote in his 1849 *History of Putnam County, N.Y.*: "The greedy, sordid, and avaricious spirit of man is making sad havoc among the beautiful mountain scenery of the Hudson. Where it will stop is more than we can tell. The sound of the ax and the railroad excavator's pick, with the steady click of the quarryman's hammer, is daily sounding in our ears, and slowly, though steadily, performing the work of demolition."[21]

Fifty years later, the destruction of river scenery was becoming widespread throughout the Hudson Valley. The forests were virtually clear cut, the historic Revolutionary War forts in the Highlands were being vandalized, and quarrying threatened to destroy the basalt cliffs of the Palisades and scar the mountain faces of the Highlands. By the turn of the century, public opinion began to crystallize, and preserving the Hudson's natural beauty became a matter of statewide interest. The Highlands, the home of the nation's leading industrialists, the place where once painters and poets had dined with manufacturers,

The old hotel at Garrison Landing.

finding sublimity in the smoke and noise of the Cold Spring Foundry, became a battleground where industry and scenery were no longer thought to be compatible. By the turn of the century, the Highlands were less fashionable, the foundry fires had ceased to burn, and the population began to decline. Those who remained began to campaign for conservation of the Highlands as wilderness. What had once been their private preserve would soon become a refuge for the middle class and the poor, a place to escape the disease ravaged city for an outing in the Highlands' mountain air. The captains of industry, Harriman, Morgan, Bigelow, Stillman and Osborn's son William Church Osborn were among the most outspoken advocates and liberal donors for this new policy.

9. The Great Outdoors: Bear Mountain Park, Conservation, and Social Welfare

FOR MUCH of the nineteenth century, nature was worshipped but was also freely sacrificed to the demands of a growing population and burgeoning commerce. By the end of the century, environmental abuse was taking its toll and the conflicts could no longer be ignored. Nationally, thundering herds of buffalo had been all but destroyed and the passenger pigeon was hunted to extinction. In the Northeast, wild game was disappearing, the commercial fisheries of the Atlantic coast had become critically depleted and the forests fell to a lumber industry which marched through the northern states "like an army of devastation," leaving the woodland soil, now exposed to wind and weather, to erode.[1] With the loss of tree cover, the soil's ability to retain rainwater and snowmelt and filter it gradually but continuously into the rivers and streams was so greatly diminished that water supplies and navigation canals were being shut down. Americans were bumping up against the reality that there were limits on the earth's capacity for renewal and those limits were being reached.

With the pace of destruction proceeding at an ever more rapid rate, the need for public action was dramatically apparent. Food, water, recreation, and commerce were all endangered by the loss of natural resources, and for a country which found great spiritual value in nature, the threats to scenery and wilderness caused no less alarm. With characteristic energy, Americans responded, and from about 1870 to 1910, at the same time that the railroad barons were developing their country estates, sweeping new initiatives were launched to assure that the public interest in natural resources would be protected, resulting in numerous new state and federal laws, new government agencies, and the formation of numerous clubs, associations, and citizen groups. During that forty year period, environmental conservation took hold in the Hudson Valley, beginning with efforts to protect the Palisades and culminating in 1907 with a public outcry for the preservation of the Highlands. For a brief but active three-year period from 1907–1910, petitions were circulated, committees formed, laws passed and repealed, millions invested, and more than 10,000 acres and two historic sites preserved.

What happened in the Highlands was part of a broader national movement which began at the close of the Civil War. At the federal level, agencies created during this era included the Game and Fisheries Commission established by Congress in 1871, and the U.S. Geological Survey formed in 1879. In 1872, the

A camping trip in the Highlands, photographed by William Thompson Howell. Courtesy of the Palisades Interstate Park Commission, Trailside Museum, Bear Mountain.

first national park was created at Yellowstone when Congress set aside two million acres in Wyoming. A federal Division of Forestry was created in 1881, and in 1891 the establishment of a system of national forests was authorized. The same year, Yosemite was added to the National Park System. In 1906, the Antiquities Act was passed, allowing areas of scientific or historical interest on federal lands to be set aside and preserved as National Monuments, resulting in the protection in 1908 of the Grand Canyon.

This growth in federal programs was matched by the formation of public interest groups such as the American Forestry Association, chartered in 1875, followed by the Appalachian Mountain Club (1876), the New York Audubon Society (1886), the Sierra Club (1892), and the National Audubon Society (1905). Resource management was so important politically to state and federal officials that President Theodore Roosevelt, himself an ardent conservationist, hosted a conference of Governors at the White House on the subject in 1908. The conference reflected a changing attitude that was widely shared among Americans—the idea that the public interest in natural resources is greater than that of the individual.

From this notion, two movements emerged, one focused on conservation of renewable resources (forests, water, game and fisheries), and the other based on the idea of preserving wilderness for its spiritual value. The roots of the preservation movement could be found in the works of nineteenth century writers, painters, and philosophers. The Knickerbocker writers, the Hudson River School painters, and their influential contemporary, Andrew Downing, made natural beauty a matter of national pride and identity, while Olmsted and Vaux, the great architects, placed on the public agenda the creation of municipal parks where the urban dweller could experience nature and the scenery first hand. Writers such as Thoreau, Burroughs, and Muir fostered the idea that the essence of the human spirit was to be found only in nature. When the western landscape paintings of Thomas Moran made their way back east, it did not require a great leap of thought to suggest that a national park be created to preserve the breathtaking splendor of Yellowstone.[2]

In the Hudson Valley, the romantic movement of the pre-Civil War era helped set the stage for preservation of a different kind. Ironically, while Osborn and his contemporaries built castles and ruins to imitate those along the Rhine, few people gave any thought to the protection of the historic forts and scenery which had given the Hudson its reputation.[3] The walls of Forts Clinton and Montgomery had crumbled and fallen unnoticed until the turn of the century when a few concerned citizens realized that irreplaceable landmarks of the nation's early history might be lost.

In 1908, William Thompson Howell, an avid hiker who knew almost every inch of the Highlands' rough terrain, began to campaign for their preservation. He wrote:

> It would seem eminently fitting . . . that the Highlands should be preserved by the national government. Not only do they contain the great military school

> and naval magazine, together with the State encampment, but two famous battlefields. . . . There are also the 'passes of the Highlands,' a phrase that occurs again and again in the letters of Washington and other Revolutionary leaders. All along the range from New Jersey northeastwardly into Connecticut, the Highlands swarm in historic associations. . . . But most interesting of all, in the remoter hills and forests are sleeping the almost forgotten, overgrown military roads, ruined mines and forges, abandoned farms and other reminders of very early days in our national life.[4]

Howell, who worked for the telephone company, also contributed articles about nature and conservation to the *Tribune* and the *Sun*. His diary, more like a scrapbook, is preserved in the archives of the New York Public Library and contains photographs he took in his Highland rambles as well as his written observations on the natural history, the local lore, and the political events which brewed in the Highlands at the turn of the century. The diary reveals a man who cared more passionately about protecting the Highlands than anything else. It includes copies of every article, every piece of proposed legislation and every public relations document which appeared about the Highlands. Written by a man who was not among the power brokers of the elite, it contains a few incisive remarks about politics in New York. Compiled at a time when, according to Howell, "Probably more has been written about the Highlands during the past ten years than in all their previous history,"[5] the diary is filled to overflowing and gives a clear picture of the politics of preservation at the turn of the century.

By the time Howell set his idea for a national historic preserve to paper, events had been set in motion in New York state and nationally to give credence to his proposal. The notion of historic preservation was first advanced by Washington Irving, a great admirer of the nation's first President who was his

William Thompson Howell photographing in the Highlands. Courtesy of the Palisades Interstate Park Commission, Trailside Museum, Bear Mountain.

namesake. It was Irving who first proposed that the Hasbrouck House in Newburgh, General Washington's headquarters in 1782–83, be given protection. When citizens of Newburgh purchased the Hasbrouck House that year and presented it to the state, it was the first occurrence in the United States of public action to preserve an historic structure for public visitation. Governor Hamilton Fish signed into law the bill accepting the historic site.

Twenty years later, Frederick Law Olmsted, creator of Central Park, began to discuss the concept of scenic resource preservation. In 1869, he proposed that the lands adjacent to Niagara Falls be made public, and in 1882, a group of leading citizens organized the Niagara Falls Association to seek enactment of a state preserve there. Through a media campaign with public speeches, pamphlets, and newspaper articles, this goal was achieved and the Niagara Reservation was dedicated by state law in 1885. The purchase of Niagara was the first time that a state had purchased property for purely scenic purposes.

The same individuals who formed the Niagara Falls Association then shifted their attention to the creation of a permanent body for the protection of scenic and historic places in New York State, and in 1895 the New York State Legislature chartered the "Trustees of Scenic and Historic Places and Objects in the State of New York" to act as a public trustee of scenic, historic, and scientific properties on behalf of the state and municipal governments and for individuals who might deed or bequeath their properties to it. The organization's name was changed in 1901 to the American Scenic and Historic Preservation Society and became New York's first agency for general conservation purposes.[6] The creation of the American Scenic and Historic Preservation Society was based on the idea that historic landmarks and scenery constitute equal parts of a national heritage.

The second great rallying point for scenic conservation in New York State was the Hudson's Palisades, and this would set the stage for the conservation of Highlands scenery a decade later. The Palisades riverfront hills provided "trap rock" of high quality at low cost, being close to ready markets and easily loaded ónto railroads or river barges. For decades, quarrying on the Palisades had been confined to the slope at the foot of the cliffs. However, in the 1890s demand increased for trap rock to be used for paving stones, for vessel ballast, and for crushed rock in concrete to build the new skyscrapers of New York and for macadam and later concrete roads. This made it profitable for quarrymen to use heavy machinery, and they moved their operations to the tops of the cliffs. As one report noted, the river reverberated "morning, noon and night . . . with heavy explosions of dynamite as one after another of the most picturesque of the columns in the stretch between Edgewater and Englewood Landing toppled into ruin in clouds of gray dust." Near the sites of the present-day George Washington Bridge, quarrymen blasted out 12,000 cubic yards per day. The American Scenic and Historic Preservation Society observed, "The effect was not only to destroy the perpendicular face of the great ramparts but, owing to the shallow nature of the west side of the River, to create a fill upon which industrial plants might easily have been built."[7]

The Palisades.

Public sentiment began to crystallize in opposition to the destruction of the Palisades, and in 1895, the states of New York and New Jersey sought to have the federal government make a military reservation and national park of the Palisades. Apparently, this was modeled after the German practice of acquiring large tracts of land for troop maneuvers. However, the House of Representatives Committee on Military Affairs of the U.S. rejected this proposal in 1897 and again in 1898 on the grounds that the Palisades had no military value, and the blasting continued.

While this debate in Congress took place, the New Jersey State Federation of Women's Clubs met in Englewood in 1897 to devise an alternative plan. The turn of the the century was a time of social expression for women, especially women of means, and the Hudson Valley was a center of activity, where the concept of progressive social reform acquired a new dimension as women became the driving force behind efforts to protect the natural, historic and cultural heritage of the region. Through groups such as the Federation of Women's Clubs, they became active members of the conservation movement. Several also played important individual leadership roles. Where they were less likely to be personally successful, they brought in their influential husbands.

The New Jersey Federation became very concerned about the need to protect the Palisades from quarrying and sent a delegation to meet with Governor Foster Voorhies of New Jersey to propose the creation of a joint New York-New Jersey commission to study opportunities to protect the Palisades and report on the possibility of joint action by the two states. Andrew H. Green, known as the "Father of Greater New York" for his role in consolidating into one the cities of New York and Brooklyn along with outlying communities, was also a founder of the American Scenic and Historic Preservation Society, and he approached New York State Governor Theodore Roosevelt with a similar proposal. In 1899, after Congress had rejected the military reservation idea, a commission was created to work on a solution to the problems involved in the creation and management of an interstate park. No precedent for such a body existed, and the situation was made more difficult by the fact that the Palisades were located primarily in New Jersey, but their preservation primarily benefited the people of New York who viewed the Palisades from the opposite shore. Since New Jersey was unlikely to give up its land to New York, and New Yorkers were unlikely to pay for park land in New Jersey, a creative solution was needed.

The Study Commission proposed the creation of a ten-member permanent Palisades Interstate Park Commission (PIPC) with five members each to be appointed by the Governors of New York and New Jersey and a commission president alternating from each state. The Commission would have the power to acquire whatever land was necessary to protect the Palisades. Two thirds of the costs of operating the Commission would be paid by New York and one third by New Jersey.[8]

In 1900 bills were introduced in both state legislatures. In New York, the legislation passed easily. In New Jersey, quarrying interests lobbied against the bill, but the New Jersey Federation of Women's Clubs mounted a spirited counter effort, and the bill was finally passed there as well. George W. Perkins, first Vice-President at New York Life Insurance Company and a trustee of American Scenic, was elected President of the New York Commission and Edwin A. Stevens, of the Stevens Institute of Technology, became President of the New Jersey Commission. In its first ten years of operation, about thirteen miles of the Palisades were preserved at a cost of $200–500 per acre and a total outlay of about $1 million. Of this amount New Jersey paid $50,000 and New York paid $410,000, all of it for the protection of land in New Jersey.

George Perkins, who was known for his ability to inspire philanthropy from men and women of wealth, secured gifts of property and money to complete the purchases, including $122,500 from J. P. Morgan (who was a trustee of American Scenic), $125,000 from Mr. and Mrs. Hamilton Twombly (she was a granddaughter of Commodore Vanderbilt), and several smaller donations including $12,000 of his own money and donations from club women of New York and New Jersey. It is said that after Perkins approached J. P. Morgan and requested the donation, the financier not only agreed to the gift but actually invited Perkins to join his firm, which Perkins accepted.[9] In time, he was to become known as Morgan's "right hand man." Although New Jersey paid little

George W. Perkins.

for the preservation of the land, it gave up the commercial and tax values of the property. Constitutional issues concerning the expenditure of New York State funds to buy land in New Jersey were never raised.

The PIPC proved to be quite successful, and rapidly halted the quarrying which had threatened to destroy the Palisades forever. In 1906, the jurisdiction of the Commission was extended further into New York to include Hook Mountain, where quarrying was continuing at a furious pace, and Stony Point, the site of one of the major Revolutionary War battlefields.

Thus, the jurisdiction of the Palisades Interstate Park Commission extended up to the Highlands but did not include them, with predictable results. In a letter to Eben E. Olcott, a man named F. P. Albert, an ardent conservationist, wrote "the stoppage of quarrying operations along the Palisades cliffs has had the effect of driving the quarryman to the nearest highlands above, and to spread rather than check the disaster."[10] Recognizing this threat, in 1907, the American Scenic and Historic Preservation Society turned its attention to the Hudson Highlands and began to discuss the possibility of state or federal legislation to protect the region. With the waning and eventual demise of the iron foundry and the closing of a chemical factory near Anthony's Nose, the population of the Highlands had declined to about a quarter of its Civil War era levels. In the words of William Thompson Howell: "The picturesque, quaint and altogether delightful village of Cold Spring had no industry of any importance, and is in fact the deadest town on the lower Hudson."[11] The Highlands, which had undergone a development boom for fifty years or more, were reverting to wilderness at a time when the public was in a frame of mind to protect the region's heritage of historic sites and scenery.

It seemed to members of American Scenic that federal legislation would be most appropriate in view of the Highlands' role in the American Revolution. The federal government also had a deeper pocket; federal protection would "relieve the state of a financial burden," society members concluded. Reflecting the prevailing concerns of the turn of the century, the purposes of the legislation as identified by American Scenic would be:

- protection of natural beauty
- preservation of historic landmarks
- prevention of further pollution of the river; and
- protection of the forest cover for timber supply and watershed protection.[12]

The group drafted a bill which would create a park of sixty-five square miles on the east side of the river and fifty-seven on the west side of the river between Peekskill Bay on the south and Newburgh Bay on the north. Because of the wild and mountainous character of the terrain and the "general absence of cultivation or improvements," it was thought that the property could be acquired for "a reasonable appropriation," based on the stumpage value of the timber. The Society suggested that some property owners might be convinced to sell what today we know as conservation easements, which would allow them the continued use and ownership of their land but would prohibit forever any uses which might destroy the scenery.

One of the most vocal supporters of the preservation of the Highlands was Dr. Edward Partridge, a Trustee of American Scenic and a resident of Cornwall. Dr. Partridge wrote numerous articles on the topic for newspapers and magazines. This excerpt from "The Outlook" on November 9, 1907, was typical:

> The Hudson river presents, throughout the fifteen or twenty miles of its course in the Highlands, its most picturesque and boldly beautiful section. Every American, from near or remote parts of these United States, is influenced to

> visit the great Hudson, and here his eye is pleased and his patriotic feeling is strengthened as he realizes the difficulties, the hardships, and courage of the founders of this great Republic. . . . This region, created into a National preserve, would serve as a most suitable memorial of the war which yielded to us our National independence.[13]

In other writings, Partridge reminded his audiences of the Highlands' role in literary traditions: "This region has found its place in American fiction through Cooper and Irving, while its charms have been told by Morris and Bryant and by the lamented Drake in the "Culprit Fay."[14] Partridge proposed that the National Park be governed by a commission created with powers to prevent disfiguring and offensive industries such as "iron works, chemical works, etc." He proposed the implementation of scientific forest management and the acquisition of conservation easements.

Partridge was a prominent New York City obstetrician at the Nursery and Child's Hospital. He had purchased a large tract of land on Storm King Mountain where he built a beautiful estate. The property offered a sweeping northwest view of Newburgh Bay, the Shawangunk Mountains and the Catskills. From his home in the Highlands he came to develop a deep love of the region's beauty and history. He was a friend of William Thompson Howell and Mr. and Mrs. E. H. Harriman, and he shared his concern for the preservation of the Highlands with both of them. An avid outdoorsman, he took frequent expeditions into the woods. Howell recounts that Partridge joined him on a camping trip and brought the provisions for a group of twenty people, a repast which included sardines on Uneeda biscuits, bacon and coffee.

Partridge organized some of his friends and neighbors including J. P. Morgan, William Church Osborn (son of railroader William Henry Osborn), General Roe of West Point, Herbert Satterlee of Fort Montgomery, Dr. George Kunz, Lyman Abbott and others to seek the creation of a National Park in the Highlands.[15] Together, in 1907, they formed the Association for the Protection of the Highlands of the Hudson to carry forward the campaign. It was through Partridge's efforts that the idea was also taken up by the American Scenic and Historic Preservation Society.

As the two organizations pressed their case in Washington, they were joined by individuals such as William Thompson Howell, who publicized the idea in articles to the *Tribune* and the *Sun*, and F. P. Albert, who conducted a tireless letter-writing campaign.

In the summer of 1907, a petition was drawn up by a Mr. Bush-Brown and circulated to the public at Washington's Headquarters, the Hasbrouck House in Newburgh, where it was signed by people from many states. The petition requested President Theodore Roosevelt and the Congress to create a National Park in the Highlands of the Hudson River. The petition noted the unsurpassed beauty of the region and the threat of quarrying moving in from both the north and south, and concluded "While the Federal Government has established many parks to commemorate the Civil War which was due to sectional dissensions now happily past, we have no National Park to commemorate the brave

Hudson Fulton Celebration official poster.

deeds of the Founders of the Nation which are the cherished heritage of the American people."[16] However, these efforts were unsuccessful. Authorities in Washington responded that there was nothing of spectacular importance to commemorate, and the idea had to be abandoned.

About this time, New York State was planning a commemorative celebration of Henry Hudson's 1609 voyage and Fulton's 1807 successful experiment with steam navigation. The Hudson-Fulton celebration was being organized by a legislatively created Commission, established in 1905 by New York State to prepare a series of events to take place in 1909. The Trustees of the Commission looked at all manner of ways to commemorate these two historic events. Plans ranged from parades and fireworks to construction of a replica of the *Half Moon* to the creation of parks and the protection of river scenery. Preparations for the celebration aroused a great deal of civic pride among river communities and stimulated a variety of local improvements including the creation of many municipal parks. Having been rebuffed in Washington, the American Scenic and Historic Preservation Society and the Association for the Protection of the Highlands of the Hudson turned their efforts from the federal government to the state government. Their efforts created a groundswell of support and numerous people petitioned for the Trustees to consider the establishment of a state park in the Highlands. Among the concerned citizens was Mornay Williams who appealed to Frederick Seward, chairman of the Hudson-Fulton Celebration planning committee on December 30, 1907: "I write as a citizen who has been for many years interested in all matters pertaining to the welfare, integrity and beauty of the State, . . . that the best memorial of the discovery of the Hudson would be the inauguration of a movement to save the Highlands of the Hudson from exploitation by quarrymen or other commercial manufacturing enterprise. The Highlands of the Hudson, now wooded and practically unoccupied land, could, if taken at once, be turned into a sort of public reservation."[17]

F. P. Albert, continuing his letter-writing campaign, sent similar correspondence to Francis Lynde Stetson, Chairman of the Law Committee of the Hudson-Fulton Celebration. Stetson, however, apparently found the idea unrealistic. He replied that "the expenditure of protecting the highlands, from the Tappan-Zee to Newburgh Bay would be enormous." He inquired whether Albert had any reason to believe that under present conditions the state of New York would be willing to provide a sum sufficient for the purchase of this private property. Undaunted, F.P. Albert responded: "these Headlands mark the birth of our nation; what more appropriate tribute than the preservation of these natural monuments of the Hudson?" Arguing that cost of acquisition now is cheaper than acquisition later after development has occurred, he wrote:

> The experience in saving the Palisades justifies this conclusion and warrants the belief that the work of protection will be welcomed rather than opposed by the residential interests along the river; for it must be borne in mind that quarries, in their immediate vicinity, are in some respect less objectionable than such industries as bone-boiling, chemical works, powder magazines and similar pursuits in life . . . which seek the secluded places along the shores of our rivers and bays. . . . It is also of the greatest importance in considering the feasibility of this project, to view it from an economical as well as from a sentimental standpoint. Some day public opinion will stop this defacement; the limit of public endurance will be reached and the expense of stopping this vandalism and preventing its growth will be infinitely greater than at present.[18]

On September 23, 1908, the Trustees of the Hudson Fulton Celebration gave their nod of approval and established a special committee on Hudson River Scenery; however, as F.P. Albert would later learn, Francis Lynde Stetson was not on their side.

The purpose of the Scenery Committee as reported in the Fourth Annual Report to the Legislature of the Hudson-Fulton Celebration Commission was "to promote legislation, to encourage private generosity, to foster public sentiment, and to cooperate with other organizations with a view to securing the preservation of the natural beauty of the Hudson River." Its principal goal was to preserve the Highlands. It was felt that "there could be no more fitting memorial of Hudson's and Fulton's achievements than provision that these silent witnesses of the passing events of centuries should remain undespoiled of their natural beauty and grandeur."

New York State Chief Judge Alton B. Parker, a Hudson riverfront resident in West Park and Democratic nominee for the Presidency in 1904, was appointed Chairman of the Hudson River Scenery committee and F. P. Albert was rewarded for his efforts, being appointed Secretary. Among the other members were Ambassador John Bigelow, Dr. Edward Partridge, George Perkins, Major-General Charles Roe, Colonel Herbert Satterlee, Frederick Lamb, Benjamin B. Odell, and J. DuPratt White.

The committee set to work immediately and, after a few months of work, presented a report to the Trustees seeking their endorsement of a plan to pass state legislation establishing a state park in the Highlands. They prefaced their remarks with references to the region's role in American cultural history and suggested that property owners in the area would support the efforts of the state to protect the land:

> While it is true that this Commission is "Hudson-Fulton" in name, and is concerned with providing memorials to the man who discovered the Hudson River, and who sailed past the Highlands long before the establishment of our republic; also to the man who adapted steam to navigation; yet you are preserving a region full of the memories of our Dutch traditions, described in American poetry and fiction, and replete in associations with early American history.

The Hudson River Scenery Committee went on to note that the value of quarries at Hook Mountain was $5 million, whereas before quarrying, the entire mountain could have been bought for $25 to 30,000. Reminding the Trustees that 13 miles of the Palisades had been bought for a public expenditure of $450,000, the report suggested that preservation of the Highlands would be a wise investment and could be accomplished at a reasonable price:

> it may be said that a number of property owners in this region, appreciating the effort to preserve this natural scenery, will help this movement by turning in their lands at a nominal fee. . . . To delay protection of the natural scenery of the Hudson River until after serious damage has been done, and then in preventing the further defacement to be compelled to pay enormously for damage already inflicted, does not appeal to the practical business mind as a position in which the interests of the public should be placed or remain unprovided for.[19]

The Hudson-Fulton Commission voted to support two pieces of legislation which were introduced in the state legislature in 1909: the Wainwright-Merritt bill for the protection of the wild forest growth, and the Bennett bill for the establishment of a state park "as a memorial to Henry Hudson and Robert

Fulton . . . to preserve the natural scenery of the Hudson River." The provisions of the Bennett bill designated a strip one-half-mile wide on both sides of the Hudson River from Jones Point to Storm King to be studied as a potential park. It gave the state authority to acquire lands and required that a map of lands to be acquired be drawn up. It also authorized the use of condemnation power for this purpose if necessary. The Bennett bill was the Committee's top priority.

The Wainright bill proposed creation of a State Forest Reserve in the Hudson Highlands, comprising about 75,000 square miles on each side of the river from the northern gate to the southern gate of the Highlands. The bill's purpose was to preserve the region's beauty, apply scientific forestry methods, and prevent clearcutting. Its goal would be aesthetic beauty instead of timber for profit. The Reserve was to be administered by the Forest, Fish and Game Commission (a predecessor to the Department of Environmental Conservation of today), which was given the power to manage land, appoint a forester, impose fines, and buy land and easements. It would extend the police power of the state to privately owned forests for the first time, compelling management of forest land according to modern forestry practices, which emphasized sustainability of the resource and aesthetic beauty of managed forests.

The Bennett bill was viewed as the more urgent of the two bills, as it would strike at the heart of quarry operations and prevent scenic defacement of the Highlands; however, the Forest Reserve bill fit into a growing state and national movement of forest protection. In 1885, the New York State Legislature had established a forest preserve in two other major regions of New York State, the Adirondacks and the Catskills, giving them constitutional protection as "forever wild" in 1895. Forest Commissions were being established across the country in response to problems caused by rapid deforestation of large tracts of wilderness. Flooding and loss of habitat for wildlife and game were two concerns. It was also believed that trees were filters for miasmic swamp air and should be preserved in the interest of public health.

By 1908, deforestation in the Highlands had occurred repeatedly over more than a century. Hessian descendants of Revolutionary War mercenaries stayed in the Highlands area and cut its timber to supply cordwood for conversion to charcoal to keep the iron furnaces burning. When the iron industry converted to coal after the Civil War, they began marketing cordwood to the Hudson Valley's brickyards. As American Scenic reported, "ownership was ignored largely and the forest was cut off as grass is mowed." In the more accessible spots, the lumber companies moved in to take the place of the small crop foresters. While removal of the Highlands' forest cover had not disturbed painters, poets, and tourists during the earlier part of the century, the situation had changed dramatically by 1900. The prevailing view was that "all this wide area of future forest, containing many lakes of great beauty, was being sacrificed."[20]

The Hudson River Scenery Committee received an appropriation of $250, of which it spent $229 to print and mail 15,000 illustrated pamphlets to organizations, clubs and newspapers describing the two bills. This publicity blitz resulted in nearly unanimous press endorsement of the two bills to preserve the

The cord wood industry remained active in the Highlands well into the twentieth century. Courtesy of the Palisades Interstate Park Commission, Trailside Museum, Bear Mountain.

Highlands, and the committee soon won widespread support for its efforts to create a state park and a state forest in the Highlands. The Federation of Women's Clubs of the State of New York, through their committee on Forestry, took an active interest in the cause, and the publications of the Hudson River Scenery Committee were circulated throughout the State by the clubs of that organization. As the Hudson-Fulton Committee's Annual Report to the Legislature recorded, "No feature of the Celebration received more thorough endorsement by the press of the City and State of New York than that for the preservation of the romantic scenery of the river."

However, the Bennett bill was not universally favored by residents of the Highlands. William Church Osborn opposed it, while his brother Henry Fairfield Osborn supported it. Both were ardent conservationists: Henry Fairfield Osborn was the president of the Museum of Natural History. Senator Schlosser of Fishkill and Congressman Hamilton Fish of "Glenclyffe" in Garrison, both east shore politicians, opposed it, but the latter's brother Stuyvesant Fish took no position on the measure. It was passed by the state Assembly, but in spite of heavy statewide backing, it never came to a vote in the Senate. Members of the Hudson River Scenery Committee of the Hudson-Fulton Celebration learned that Francis Lynde Stetson from Hudson-Fulton's Law and Legislation Committee had killed it.

The Wainwright Forest Preseve bill passed both houses of the legislature on May 8, 1909, but not before the east shore of the Highlands was removed from it. In the end, it designated seventy-five square miles on the west side of the Hudson as a Forest Reserve subject to forest management by the Forest, Fish and Game Commission. All in all, it was a disappointing year for the Committee, as the celebration came and went with no Hudson Highlands State Park to commemorate it. The report of the Scenery Committee to the Trustees of the Hudson-Fulton celebration stated:

> It is to be regretted that it cannot be said to the multitudes who witness your decorous and impressive ceremonies: 'The law making power of this State has said, no longer shall the ruthless defacement of the beautiful Hudson River be permitted.' But you can at least give assurance that the work will not stop until that most to be desired result is accomplished."[21]

The Forest, Fish and Game Commission appointed F. F. Moon as Forester of the Hudson Highlands Forest Reservation, and Moon spent the next year traveling to schoolhouse meetings throughout the region trying to get woodsmen to harvest trees for telephone poles instead of cordwood. This would promote long-term gain and prevent clearcutting according to sound forestry practices. However this was clearly not sufficient to protect the Highlands' scenery and the public expectation had been raised to a very high level. Pressure was building for the preservation of a disappearing heritage. In the end, the spark that ignited public action was a state proposal to build a penitentiary at Bear Mountain.

On March 15, 1909, William Thompson Howell was hiking through the woods when he came to a clearing with a sign board nailed to a tree. It read:

STATE PROPERTY
SING SING PRISON
TRESPASSERS MAY BE SHOT

A few months later the State of New York had erected a stockade there—on the terrace of land where American militia had fought the British in the famous battles of Forts Clinton and Montgomery. Within its walls were hundreds of convicts who were brought in by the government to clear the site for the relocation of Sing Sing Prison to Bear Mountain.

The desecration of this historic ground shocked New Yorkers and moved

1910 photo of Sing-Sing prison stockade Courtesy of the Palisades Interstate Park Commission, Bear Mountain.

two men to action: George W. Perkins and Edward H. Harriman. Perkins, now a banking partner of J.P. Morgan, a political ally of Theodore Roosevelt, and head of the newly formed Palisades Interstate Park Commission, had devoted the previous eight years to saving the Palisades from destruction by quarrymen. Harriman, the railroad magnate, possessed a strong interest in wilderness protection. He had extensive land holdings in the adjoining Ramapo Highlands and at his estate there, "Arden," he had spent many hours discussing conservation of the Highlands with his friend Dr. Edward Partridge.

In 1909, Perkins and Harriman devised a grand plan to preserve the entire Highlands chain, from the Ramapos to the Hudson, including the famous battlegrounds at Forts Montgomery and Clinton, and to extend the jurisdiction of the Palisades Park Commission north from Stony Point to Newburgh. Harriman presented the idea to Governor Hughes of New York in a letter dated June 1, 1909. Harriman proposed to donate nineteen square miles of his own estate in the Ramapos as well as a gift of $1 million to acquire property between it and the Hudson. His letter outlined the plan he and Perkins had envisioned: "I have thought that possibly some of the other property owners might join with me in this move," he wrote, "making it possible to secure practically the whole wild area between the Ramapo and the Hudson Rivers, extending from West Point down to below Stony Point, and again north of West Point, taking in the Crow's Nest. I feel that if this should be accomplished, the State's prison should be moved again to the other side of the river, so as not to destroy the natural beauty which can never again be replaced." He added that such a park would be of particular benefit to New York City and could be used as a health camp for sufferers of tuberculosis, where they could enjoy the outdoor life.

Governor Hughes responded soon after with a letter indicating his interest in the concept. "The creation of a public park embracing a wide expanse of this beautiful highland country, controlling a great watershed, and reserving for public uses a territory so accessible to New York City, would be of incalculable benefit to the people." He also expressed his support for "the great advantages such a reservation would present for health camps and for outdoor life, and the vast possibilities which would be opened up for those who are crowded within the great city so near at hand."[22]

Harriman died a few months later, before his plan could be executed. He left his entire seventy-million-dollar estate to his wife, and it was she who converted his vision to reality. On December 15, 1909, Mary Harriman offered a gift of land and money to New York State. She set out five conditions:

- that by January 1, 1910, (16 days later) a further $1.5 million in private subscriptions be secured;
- that construction cease on the Sing Sing Prison at Bear Mountain;
- that New York State provide a match of $2.5 million for land acquisition, road building and park purposes;
- that state law be amended to extend the jurisdiction of the Palisades Park north into the Highlands and the Ramapos;
- that New Jersey appropriate an amount determined by the Interstate Park Commission to be "its fair share."

Mary Harriman further specified that the State was to purchase river frontage to improve the park's accessibility to the inhabitants of New York City and neighboring counties who could reach the Highlands by steamboat.

All of the conditions were met, and ten months later the Harrimans' son Averell, later Governor of New York, was summoned from Yale College to present the gift of $1 million and 10,000 acres to the Park Commission. The State's $2.5 million was approved by New York voters by means of an environmental bond issue.

Securing $1.5 million in private matching funds within sixteen days is testimony to the work of George Perkins who, as President of the Palisades Park Commission, had secured promises of support from a number of prominent individuals. John D. Rockefeller Sr. and J. Pierpont Morgan each contributed a half million dollars. Morgan, who had contributed liberally to the preservation of the Palisades a decade earlier, was not only concerned for the protection of the Highlands but also had a direct interest in preventing the construction of a prison so close to his summer home. The remaining half million was donated by fourteen other subscribers, including Margaret Sage, James Stillman, William K. Vanderbilt, William Rockefeller, Henry Phipps, Helen Gould, V. Everit Macy, and George Perkins himself.

The legislation adopted in 1910 which extended the jurisdiction of the PIPC to the Highlands also repealed the Hudson River Forest Reserve. However, Dr. Edward Partridge, the leading proponent of the forest reserve concept, was appointed to the Interstate Park Commission in 1913, where he would serve until his death in 1930.[23]

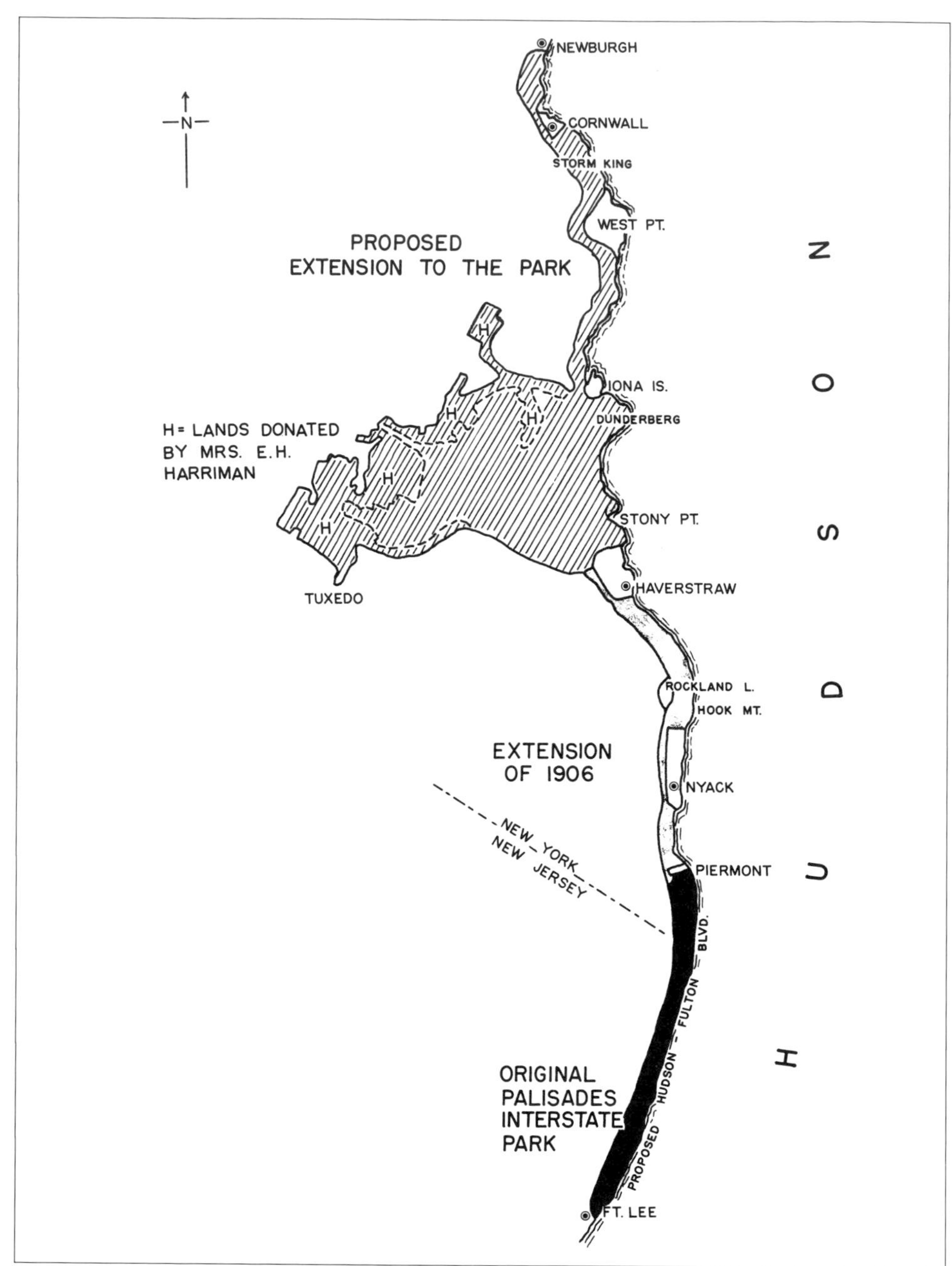

Palisades Park in 1909 with actual and proposed extensions. Courtesy of the Palisades Interstate Park Commission, Bear Mountain, reproduced with permission from *American Sublime*, by Raymond O'Brien, Columbia University Press.

The creation of the Bear Mountain-Harriman State Park fit into an emerging pattern of conservation activism by some of the nation's most prominent political, industrial, and financial leaders. Even so, it was an unprecedented act of philanthropy in terms both of its magnitude and its effect in prompting other landowners to give land or money to complete the proposed acquisition. For the Harrimans it was a natural progression in a history of conservation interest.

Although several of the railroad barons actively backed the National Park movement with an eye toward increasing passenger traffic, Harriman was one whose actions indicated more than a business interest in conservation. He and his wife Mary both loved the outdoors and maintained a friendship with the naturalist and Sierra transcendentalist John Muir. When Muir began his efforts to save the Yosemite Valley, the Harriman family came to his aid. In 1899, when Harriman sponsored an expedition to Alaska with a team of scientists, painters, and photographers, Muir was one of the group.

The Alaska trip, which was recorded in thirteen volumes containing monographs by many of the participants, offers an interesting record of Harriman's personality and his views of the natural world. The financier organized the expedition when his doctor advised him to take a vacation. Having decided to go by steamer to Alaska, where he hoped to hunt the Kodiak bear, thought to be one of the largest in the world, Harriman came up with the idea of inviting eminent scientists to join him. As if preparing an ark for a great flood, he proposed taking two zoologists, two biologists, two botanists, two geologists, and so on, to record the natural history of the subarctic. The trip ultimately included three artists, two photographers, and twenty-five scientists, as well as John Muir and naturalist John Burroughs, who wrote a narrative of the expedition.

Burroughs, who had lived in the Highlands during his early career as a teacher, described the Alaskan landscape and found it similar to that of the Highlands. He recorded the scenery of "fiords and mountain-locked bays and arms of the sea. Day after day a panorama unrolled before us with features that may have been gathered from the Highlands of the Hudson . . . with the addition of towering snow-capped mountains for background." Muir wrote of the trip: "I soon saw that Mr. Harriman was uncommon. He was taking a trip for rest, and at the same time managing his exploring guests as if we were a grateful, soothing, essential part of his rest-cure, though scientific explorers are not easily managed, and in large mixed lots are rather inflammable and explosive, especially when compressed on a ship. Nevertheless he kept us all in smooth working order; put us ashore wherever we liked, in all sorts of places—bays, coves, the mouths of streams, etc . . ." Once, at a dinner on the expedition, Muir joked about Harriman, who was not present, "I don't think Mr. Harriman is very rich. He has not as much money as I have. I have all I want and Mr. Harriman has not." Harriman got word of this remark and told Muir, "I never cared for money except as power for work. . . . What I enjoy most is the power of creation, getting into partnership with Nature and doing good, helping to feed man and beast, and making everybody and everything a little better and happier."[24]

On Wall Street, Harriman had quite a different reputation. Biographer George Kennan described Harriman as "cold, reticent and austere" when dealing with business, and cites the comment of a journalist who said "He would be a wonderful character if only he had a heart."[25] However, at home Harriman was known to be kind and sympathetic, a side of his personality revealed to scientists on the Alaska trip as well. John Muir clearly admired Harriman, describing him as a man who:

> fairly revelled in heavy dynamical work and went about it naturally and unweariedly like glaciers making landscapes, cutting canyons through ridges, carrying off hills, laying rails and bridges over lakes and rivers, mountains and plains. . . . Fortunes grew along his railroads like natural fruits. Almost everything he touched sprang up into new forms, changing the face of the whole country. . . . When big business plans were growing in his head he looked severe, with scarce a trace of the loving kindness that, like hidden radium or the deep-buried fires of ice-clad volcanoes, was ever-glowing in his heart. . . . [N]one who came nigh him could fail to feel his kindness, especially in his home, radiating a delightful peaceful atmosphere, the finest domestic weather imaginable.

In Kodiak, after a two-day trek with a Russian guide, Harriman shot his Kodiak bear and a half-grown cub. On his return, he was toasted by his "scientific associates upon having obtained one of the objects of his ambition, and upon having made at the same time an important addition to the collections of the expedition."[26]

It was a twist of fate that brought the Harrimans to the neighboring western Highlands. In 1885 their friends the Parrotts sold their Orange County estate at auction. Harriman was in attendance. With 7800 acres of standing timber, it was eagerly sought by representatives of lumber companies. Harriman, shocked at the thought of what the lumber companies would do to the land, determined on the spot to buy it for a country estate. Thus, the Harriman family established their summer residence in the Ramapo Hills, as the western Highlands are known. There, Harriman operated a dairy farm where he "amused himself by giving personal attention to the details of farming and dairying, and pioneered the development of scientific methods for improving the quality of the milk supply." He also took an interest in forestry, "personally selecting the trees, or the kinds of trees, to be preserved."[27] He also invited students from the Yale School of Forestry in New Haven, then recently established by forestry pioneer Gifford Pinchot, to spend a summer at "Arden" studying forestry methods and getting practical experience.

In 1905, Harriman decided to build a house on his property in the Ramapos and selected a mountain top 1300 feet above sea level. To make construction possible, he began by building a cable railway to the summit. The railway was capable of lifting an automobile up to the mountain top in about ten minutes. "Arden House," when completed, was made of local granite and contained only American materials in its architecture and decor. The Harrimans expanded their land holdings in the following decade to an expanse of about thirty square miles composed of about forty farms and wooded tracts. The original tract purchased from the Parrotts was already "two to three times as much land as he really needed," but he expanded his holdings to "save the beautiful wooded hills on the west side of the Hudson River from falling into the hands of timber speculators and lumbermen and preserve them intact for the use and benefit of the public."[28] Parts of this package of land formed much of the initial gift to the state.

The Harrimans' plan did not stop at mere preservation of the land. They envisioned a park which would provide fresh air recreation and camping for the urban poor. The idea proved very successful and Bear Mountain Park rapidly developed on a grand scale. By 1914, it was estimated that over one million people were streaming through the park's gates annually. Three years later the number of visitors doubled to two million. By 1919, there were twenty-nine tent settlements used by more than 50,000 people, mainly women and children. The average stay was eight days. The park also provided camping sites for 1600 Boy Scouts.[29] Bear Mountain was immensely popular and remains so to this day. It receives more visitors annually than does Yellowstone National Park.

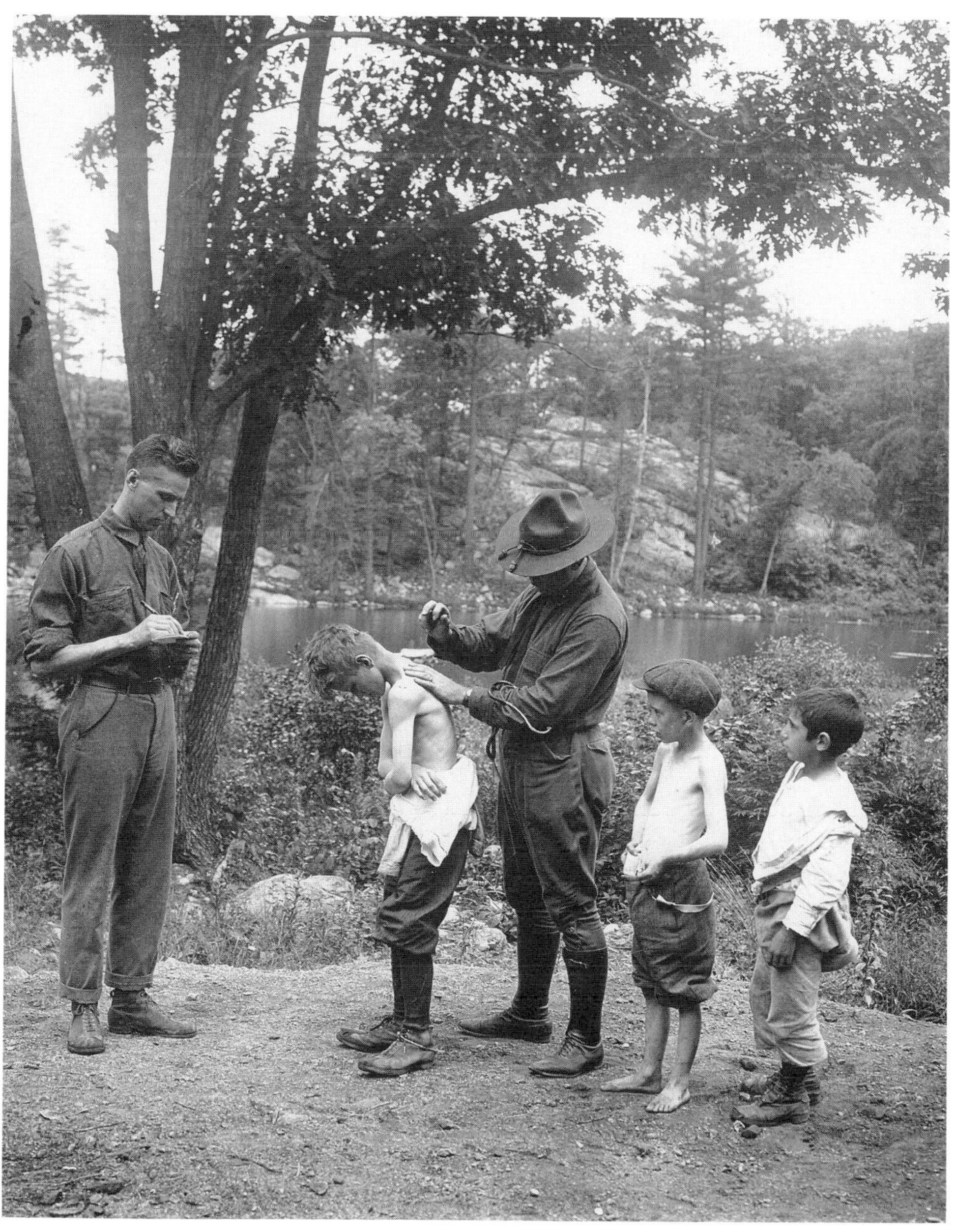

Rangers give boys a physical at Bear Mountain Park. Courtesy of the Palisades Interstate Park Commission, Bear Mountain.

Over the years, huge sums of money were invested by the States of New York and New Jersey and by private donors to expand the park. In 1929 the Harrimans' goal of a public preserve between the Ramapos and the Hudson was achieved, as the last parcels of intervening land were acquired. In later years, the remains of Revolutionary War strongholds were excavated and preserved, and on the site of Fort Clinton—with support the Rockefellers—a historical museum was erected.

To accommodate the visitors to Bear Mountain Park, the Palisades Interstate Park Commission began a vast construction program which included steamboat docks, roads, a bus shelter, and a variety of buildings and lodges. Everything was built of natural materials obtained from the park: stones from old fence rows and wood from the forest, primarily chestnut, which was rapidly dying of blight.

The park structures were designed to complement the natural surroundings. Instead of the Gothic style of architecture so popular in the nineteenth century, park planners adopted the rustic style. The best example was the Bear Mountain Inn, built in 1915, described as "a rugged heap of boulders and huge chestnut logs assembled by the hand of man, and yet following lines of such natural proportions as to resemble the eternal Hills themselves."[30] Originally it was an open air restaurant, with a luncheon counter on the first floor and a full dining room on the second. The interior is built of chestnut wood as well and was designed in the architectural style of the Adirondack "Great Camps." No windows or doors interfered with the experience of the outdoors at the Inn, although provisions were made to glass in the second story in spring and fall. Dining was said to be a culinary experience par excellence rivaling any in New York City.

As the park was undergoing this development, a grand new idea was developing in the mind of Benton MacKaye, a U.S. Labor Department policymaker charged with creating jobs through better use of natural resources. In

Bus motorcade at Bear Mountain in the early 1920's. Courtesy of the Palisades Interstate Park Commission, Bear Mountain.

1921, MacKaye, an avid hiker, was concerned about the negative effects of urbanization on the human spirit and advocated enjoyment of the outdoors as a means of spiritual regeneration. He teamed up with two friends at the American Institute of Architects who shared his philosophy—Clarence Stein, chairman of the committe on community planning and Charles Whitaker editor of the AIA *Journal*, and mapped out the idea for a footpath along the entire Appalachian Mountain chain from Maine to Georgia. The concept was published in an article by MacKaye in the AIA *Journal* and captured the imagination of hiking groups, government agencies, and scouting organizations along the entire proposed route. MacKaye found a particularly enthusiastic audience at Bear Mountain State Park, where Major William A. Welch was General Manager. Welch who was already working on a trail system for the park, began to organize a new project to create a section of the proposed "Appalachian Trail" at Bear Mountain which would take hikers in the direction of the Delaware Water Gap to the south. The Bear Mountain section of the Trail was the first created. It offically opened on October 7, 1923 and would be the pattern for other trail sections. Each would be developed independently by local and regional groups while linking with the next section according to the plan mapped out by MacKaye. In 1925, a corporation was formed to complete the trail along its entire 2025 mile length, and in 1937 this was finally accomplished.

Today, the Appalachian Trail has achieved official state and federal protection. However, for its first thirty years, it was devised solely through cooperative agreements among the many landholders along its route. In 1984, the National Park Service assumed responsibility for its management, yet it was built by volunteers and continues to be maintained by the volunteer Appalachian Trail Conference.

Bear Mountain Park and the Appalachian Trail were the creations of a time when interest in conservation and historic preservation coincided with concern for social welfare. The region's most prominent businessmen and their wives, leading government officials, and numerous citizens joined together to create the park and the trail as part of a broader social vision. During the early decades of the twentieth century, many other properties in the Hudson Highlands would be preserved by this same sector of society who found ways to protect the region's heritage while also meeting their goals for social welfare.

Much of their attention centered on the great estates, which were beginning to fall from fashion. The larger ones were white elephants, costly to heat and maintain, and requiring many servants. In many cases, the heirs did not want the responsibility or could not support the burden of expenses, and they bequeathed their family estates to institutions which would adapt them for their uses. This was a period when social welfare institutions were first coming into existence. Philanthropists were endowing them, but they needed facilities. The mission of many of these organizations was to help the poor and indigent of the city, and it was seen as a great benefit to get them into the country, where the air was fresh and the temptations for vice were few. The estates of the Highlands

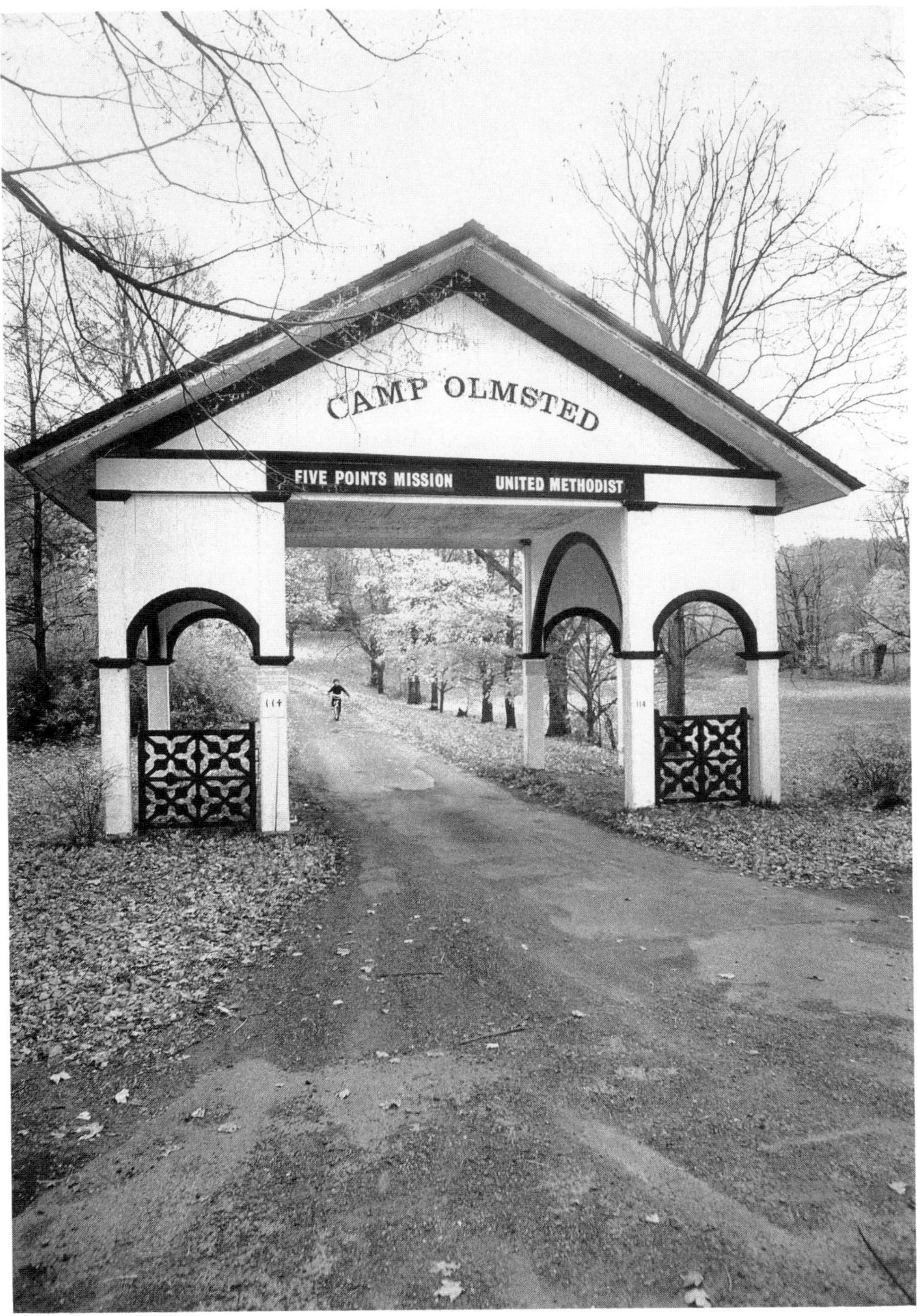

Camp Olmsted established as a fresh air camp by Rhoda Olmsted. Now a methodist retreat. Copyright © 1990 by Robert Beckhard.

offered ready-made facilities in an area known for its healthful environment, and several of them were turned over to church organizations or other institutions. Typical was Camp Olmsted, a sprawling compound in Cornwall which was purchased in 1901 by Mrs. Rhoda Olmsted to make a home in the country for children, mothers, and babies from city tenements. Over the next forty years the Jacob Ruppert estate, the Toucey estate, and the Moore house were also purchased for similar purposes.

Another woman active in conservation and philanthropy in the Highlands was Margaret Slocum Sage. At about the same time that Mr. and Mrs. Harriman were conducting discussions with Governor Hughes to stop the proposed penitentiary at Fort Clinton, Mrs. Sage came to the rescue of another Revolutionary War site in the Highlands—Constitution Island, where the American Army had encamped. Known as the Parade Ground, the Island had also served as the anchor for the iron chain which was stretched across the Hudson from West Point. Jutting into the river at World's End, the rocky island creates a bend in the river's course and the narrow passage which lent such strategic importance to that part of the Highlands during the Revolutionary War. In addition to its importance as the parade ground for Washington's army, the Island is in the foreground of the scenic view from West Point.

Equally significant, Susan and Anna Warner, authors of the bestselling novels *Hills of the Shatemuck*, and *Wide, Wide World* still resided there as they had since 1834. Although they lived on the edge of poverty and toward the end had to sell family possessions to provide for themselves, the sisters became something of an institution in the Valley. When Susan Warner died, Anna Warner offered the 280-acre island to the federal government in return for a sufficient sum to provide for her remaining years. However, despite the intervention of Elihu Root, Secretary of War, and his successor William H. Taft, Congress refused to appropriate the necessary funds. Developers offered Anna Warner almost twice as much money, but she held off, still hoping to be able to

Constitution Island Marsh. Courtesy of Scenic Hudson.

preserve the property. Later, a group of promoters proposed an amusement park for the island. At this point, Mrs. Sage intervened, purchasing Constitution Island for Anna's asking price. After considerable effort, she persuaded the federal government to accept it as an addition to West Point subject to the condition that it may never be used for another purpose. The Warner House is now open to visitors and is preserved as the sisters left it.

Mrs. Sage was a noted philanthropist and activist in support of women's suffrage and education for the self-supporting woman. An article she wrote entitled "Opportunities and Responsibilities of Leisured Women" set forth her philosophy. She believed in helping the unfortunate by providing them with a good environment, opportunity for self-support and individual responsibility, and protection from the unscrupulous. She was the second wife of Russell Sage, who built his fortune in banking, railroads, and money-lending. When her childless, misanthropic husband died in 1906, he left her $63 million in his will, which she began to distribute to charities. The Russell Sage Foundation, which she devised and ran, was the prototype of the modern American charitable foundation. This ex-schoolteacher became one of the most influential women in our history. Among her conservation-related activities were the creation of a 70,000 acre bird refuge on Marsh Island in Louisiana and the purchase of Constitution Island as a gift to the federal government.

The terms of the gift provided for Miss Warner to remain there until her death. When Anna Warner was ailing, the cadets created the Constitution Island Association to manage the property. It was the first non-profit organization in the country devoted equally to historic preservation, conservation of nature, and protection of Revolutionary War fortifications. In 1915, Anna Warner died, taking her place next to her sister in the West Point cemetery overlooking the Hudson and Constitution Island.

The actions of the Harrimans, Mrs. Sage, and those who joined them through gifts of land or money or through the power of their passions and their pens—Perkins, Partridge, Howell, Bigelow, Olmsted, Morgan, Rockefeller, Stillman, Osborn, Albert and many others—preserved a great deal of the Hudson Highlands scenery in public ownership. Their actions marked the culmination of about twenty years of efforts to preserve the Hudson's shoreline, first in the Palisades and then extending into the Highlands. During this time, nature and scenery were embraced as a vital part of our historic heritage, and a new institution to protect them, for which there was no legal precedent, was created in the form of the Palisades Interstate Park Commission. Vast sums of public and private money were spent, leaving for future generations a legacy of lands which today would cost not millions but billions to purchase for the public domain.

While the efforts of these men and women set the stage for continued public spending to support expansion of the Interstate Park, it would be decades before public interest in conservation would reach such levels again. In 1907, as the ideas of a national park, a state park, and a state forest reserve in the Highlands were being debated, plans were being drawn which would usher in a new era in the Highlands—a time when technology would flex its mighty muscle. The Hudson-Fulton celebration, which gave so much impetus to the movement to preserve the Highlands, also marked a shift in attitudes toward resource consumption. It was a turning point, a celebration of both the old and the new.

10. Highways, Bridges, and Tunnels

THE HUDSON-FULTON Celebration was as grand as any ever held. The Commission whose membership included more than eight hundred citizens and public officials spent more than four years preparing for it at a cost of over a million dollars. The festivities, lasting from September 25 to October 9, 1909, were a non-stop succession of parades, fireworks, concerts, art shows, historical and art exhibits, light shows, and boat races. The numerous ceremonies included public lectures, receptions for foreign guests, and the dedication of parks and monuments. It was as much a celebration of New York's scenery as its history, and a major feature of the event was the official dedication of the Palisades Interstate Park, which in the week before the celebration, with the donation of 60 acres of land by Mr. and Mrs. Hamilton Twombly, finally encompassed the entire waterfront of the Palisades in New York and New Jersey. In preparation for the celebration, critical momentum had also been given to the establishment of state and local parks at Bear Mountain, Inwood Hill in Manhattan, and Verplank's Point near Peekskill, as well as dozens of smaller municipal parks.

The focal point for the two-week festival was a procession from New York to Albany of full-sized replicas of Hudson's *Half Moon* and Fulton's first steamboat *Clermont* which had been built especially for the celebration. Research on the original plans and design of the *Half Moon* had been conducted by the Dutch government which organized a companion Netherlands' Hudson-Fulton Celebration Commission and built the eighty-ton replica at the Royal Ship Yards in Amsterdam as a gift to the State of New York. The replica of the *Clermont* was built in the United States by the Staten Island Ship Building Company at a cost of $15,865.15, paid for by the New York-based Celebration Commission. At 1:15 on September 25, under the eyes of more than a million spectators, the *Half Moon* sailed and the *Clermont* steamed through the harbor to a 21 gun salute leading the naval inaugural parade of the Celebration. They were followed by five squadrons totaling 800 boats—yachts, steamers, tugs, tenders, gunships, submarines, and other vessels of U.S. and foreign fleets until they reached the Official Landing at 110th St. in Manhattan, where their crews, dressed in period costume, came ashore and exchanged presentations with the Hudson-Fulton Commission. The Dutch, crewing on the *Half Moon* and their captain, dressed as Henry Hudson, presented an illuminated manuscript by which the *Half Moon* was bestowed to the State of New York; the Reverend C. S. Bullock, representing Robert Fulton, gave a short speech about the *Clermont*, and then a delegation of Japanese residents of New York presented a gift of 2100 cherry trees from Japan to be planted along Riverside Drive.

If the ceremony came off without a hitch, the preparations were not without incident. At 10:30 that morning, as the two boats left their anchorages to join the squadron, the *Clermont* slipped a set-screw and came to a stop. The *Half Moon* bore down on the disabled ship and, unable to maneuver, rammed the *Clermont* near the boiler. The *Half Moon's* anchor locked into the *Clermont's* sail, and while

Warships outlined in light and fireworks during the 1909 Hudson Fulton Celebration. New York State Hudson-Fulton Celebration Commission.

the tugboat *Dalzelline* assisted in separating the two vessels it collided with the main boom of the *Half Moon*. Apparently, however, the collision "caused more amusement than damage," and by the time of the opening ceremonies, both ships were seaworthy. After the inaugural parade in New York harbor, the *Clermont* and the *Half Moon* headed upriver for a series of community festivals timed to coincide with their arrival. The ships arrived in Albany on October 9 and in Cohoes (the northernmost point of navigation on the tidal Hudson) on October 11 for the close of the Celebration.

The waterborne parade of the *Half Moon* and the *Clermont* was accompanied by land-based events which commemorated other milestones in New York State history, including achievements in technology and civil engineering. Among the highlights was a demonstration of airplane flight by Wilbur Wright, who amazed the crowd with four flights over the New York harbor, the longest of which was an incredible 33 minutes and 33 seconds long, covering a distance of about 20 miles at an average altitude of 200 feet. To demonstrate advances in the use of electrical power, the Committee on Illuminations put on spectacular light shows which included nightly "chromatic illumination" along Riverside Drive, produced by the "Ryan Scintillator." The committee also arranged for the masts and waterlines of the international sailing fleet assembled in the harbor to be strung with thousands of sparkling globes, which by pre-arranged signal would be lighted at eight o'clock. A historical pageant comprised of fifty-three horse-drawn floats featured public works achievements such as the construction of the Erie Canal and the Croton public water supply system.[1]

Looking back, the Hudson-Fulton Celebration marked the high point in public appreciation of the river's history and scenery as well as a shift in the public mood, when enthusiasm for technology would dominate the interests of citizens and government. It also marked the end of an era when scenery and industry were celebrated simultaneously and without conflict. Locally and nationally, these values began to clash. In California, a dramatic battle pitted proponents of wilderness preservation against advocates of a utilitarian view of natural resources. The decision in 1913 to dam the Tuolumne River where it flowed through the spectacular Hetch-Hetchy gorge in Yosemite National Park in order to create a public water supply for San Francisco touched off a national debate involving every member of Congress. Although the dam was built, the idea that a wilderness river valley might have equal or greater value to humanity in its unspoiled state than dammed and filled as a reservoir was gaining acceptance in American life, setting the stage for future conflicts over the appropriate use of natural resources.

Similar forces were at work in the Highlands, when the desire to preserve nature gave way to the need to harness it for public works, but unlike Hetch-Hetchy, the shift in priorities caused a ripple but not a stir. In the Highlands the clash would come later. For a two-decade period beginning at the time of the Hudson Fulton Celebration, these mountains were once again sketched on drawing tablets—this time not by artists but by draftsmen—as engineers, equipped with modern machinery and materials, prepared to test the mountains' granite resistance. The Highlands—once a bar to development—now became the location of three of the most significant engineering achievements of the time. From 1907 to 1924, the rumble of dynamite and the drumming of hammer drills could be heard from Storm King to Anthony's Nose, as previously impenetrable mountain terrain was challenged and conquered.

However the Highlands did not yield easily, even to modern science and technology, and attempts to breach the Highlands' barrier seemed at times impossible. Once again public fear and superstition was stirred. Ultimately successful, each project affected the scenery of the Highlands. The Storm King Highway left a lasting scar on the Highlands' Northern Gate; the Bear Mountain Bridge, accused of being "an infliction of ugliness" when first proposed, now graces the Highlands as a work of monumental beauty; the tunnel of the Catskill Aqueduct, the most difficult project of all, left little visible imprint despite its massive effects. Ironically, the promoters of these projects were, in some cases, the same people who had spoken so effectively for the preservation of the Highlands' scenery, and in the case of the bridge and the highway, the newly created parks and scenery provided justification for their development. The clash between the use of natural resources and the preservation of wilderness scenery—so dramatically evident at Hetch-Hetchy—was barely felt in the Highlands where the enthusiasm for technologiy captured the imagination of the public and their taste-makers.

Of the three Hudson Highlands undertakings, the aqueduct tunnel best illustrates the engineering advances characteristic of this period and the changing attitudes which accompanied the new technology. The need for each of these projects was driven by New York's expanding population, which was growing at the rate of one million per decade. In the case of the aqueduct, the driving force was the need to supply New York City with water. It was one of the most ambitious water supply systems ever built, and the tunnel crossing in the Highlands was the most difficult engineering feat of the entire project.

The Catskill Aqueduct was conceived in 1905 as a grand plan to flood twenty-four square miles of Catskill Mountain watershed. It would be the longest and most expensive aqueduct in the world. A twenty-year project, it would tap the watersheds of two Catskill Mountain streams, the Esopus and Schoharie, collecting their waters in two reservoirs. These reservoirs would feed into a central aqueduct system. The construction was to be accomplished in two stages. As wide as a subway tunnel, the aqueduct would have a capacity in excess of 500 million gallons per day (m.g.d.), with 250 m.g.d.capacity being supplied by each of the two watersheds.

The aqueduct would be designed to operate on the same principle utilized by the Romans thousands of years ago. Starting in the high ground of the Catskills, it would be constructed on a gradient so that the water would flow ever so slightly downhill until it reached New York. The water pressure would be such that no pumps would be required for the entire length. The Ashokan Reservoir, when filled, would reach an elevation of 590 feet above sea level. Its destination, the Hill View Reservoir on the northern border of New York, would be 300 feet lower than the Ashokan, an elevation sufficient to permit gravity flow and a self-delivering system.

The plan won approval, and by the end of 1905, New York's Board of Water Supply had begun mapping out the project in specific detail. On its route to New York, the aqueduct would have to pass through tunnels, around and through hills, and across valleys, using a variety of construction methods. It would follow the contours of the land, but never rise above the hydraulic gradient, since this would require pumping. Tunnels would have to be driven through hills and mountains which could not conveniently be circumvented by the earth-covered concrete trenches used for most of the system.

However, it was impossible to follow land forms all the way to the city. At some point, stream valleys and the Hudson River had to be crossed. In Roman times, the solution would have been to build multi-arched overland bridges to carry the water at its proper gradient. However, advances in mechanical power and the use of dynamite made it possible to construct tunnels under these stream and river valleys, using water pressure to bring the water back up to the appropriate level on the other side. This technique was to be tried for the first time in the construction of the aqueduct.

The engineering of the aqueduct depended directly on an accurate analysis of the geology of the Catskills and the Hudson Valley. A team of trained geologists, headed by Charles Berkey, was assembled to assist in determining the least costly route. Fault zones, as well as bedrock and channel conditions, ruled out all choices but one—a Storm King-Breakneck crossing—where a band of granite rock fairly uniform in quality and comparatively free of groundwater intrusion was found to extend across the river valley. On the advice of the geologists, the engineering team drew up plans for a U-shaped tunnel leaving the gradient near Cornwall at an elevation of 418 feet above sea level, dropping vertically below the surface of the Hudson, crossing through solid rock under the gorge, and resurfacing on Breakneck Ridge north of Cold Spring at the gradient—395 feet above the river's surface.[2] Operating as an inverted siphon, water pressure would power the rise up Breakneck, and from there the downhill flow would begin again. The tunnel would have to be buried at least 150 feet deep in solid rock to resist a bursting pressure estimated at 480 pounds per square inch.

Groundbreaking ceremonies for the aqueduct were held near Cold Spring on Thursday June 20, 1907. New York City's Mayor McClellan, using a silver spade, cut out a square of sod from the Newell Farm, where the first two acres of land for the aqueduct had been acquired by the City, and declared the project begun. In a full-page Sunday feature on June 23, *The New York Times* reported on the difficulties which would be encountered in constructing the aqueduct system and offered the following observations about the Storm King crossing: "of this gigantic undertaking . . . the dip under the Hudson is the most striking of the numerous engineering problems. . . . No such tunnel or siphon has ever been built, yet the engineers are confident of the success of their plans, drawing on experience in the constructing of mine tunnels, a science which has developed very rapidly in recent years."

Work on the Storm King shaft had actually started nearly four months before the groundbreaking, on February 23, 1907. The first problem was to determine the depth to bedrock of the granite seam underlying the river channel. Using drills mounted on pile-driving scows, kept in position by ten anchors each weighing two to three tons, contractors bored through the muck, gravel, and boulders of the river channel in an attempt to find solid bottom. Division Engineer Lazarus White described it as the most difficult task involved in the construction of the entire aqueduct system:

"The severe gales and rough water encountered at times make it hard to keep the scow exactly in position, while any movement, if it is toward the hole, is likely to bend the standing casing and destroy the borings." White added that "No work known to the writer requires a greater degree of patience, ingenuity and perseverance than boring through hard surface under the conditions prevailing at the Storm King crossing. In addition to swift tides and heavy river traffic, various governmental restrictions as the space taken, etc., there is a short working season, entirely too short to put down a 700-foot hole. This made it necessary to disconnect at the river bottom and cap the hole, recovering it if possible in the spring. . . . When at last one hole was supposed to have reached bottom at the depth of about six hundred feet, a river steamer which had broken a propeller drifted on the boring rig, bending the casing when only a few feet of core had been pulled."[3]

The Proceedings of Municipal Engineers for 1910 reported that "Of fifteen river holes, six reached rock it is believed; of the other nine, two were knocked over by passing vessels, one by its own scow, one fell over of its own weight, two were abandoned because of conditions at their bottoms, two were discontinued because supplanted in usefulness by other holes, and one is still in progress."[4]

Four years later, the bedrock of the Hudson had yet to be reached, and public confidence in the project began to flag. White reported that "The difficulties of the boring came to be associated with the entire work and people began to believe the river bottomless." Rumors circulated that the river contained an impenetrable rift, and according to another history of the project, the *World* published a cartoon "showing a man on top of Storm King squirting the aqueduct water through a hose to a man on top of Breakneck who caught it in a tub."[5] At one point, the *New York Times* reported that the water supply commission intended to change the route of the crossing to the Newburgh area. This would require a tunnel four to five times longer, but would be necessary because engineers could find no solid rock at the site of the proposed Storm King crossing.

Despite all adversity, the crews persisted in their explorations, and the results of the drilling profiles at Storm King and the other six Hudson River locations provided sufficient information for geologist Berkey to guess at the likely depth of the channel. Work on the aqueduct tunnel provided the first opportunity to study the subsurface geology of the Highlands gorge. The various resulting hypotheses on its geologic history would lead to vastly different conclusions about the shape, depth, and profile of the bedrock underlying the

Drilling for bedrock. New York City Board of Water Supply.

Storm King gorge. The drilling showed that the Hudson through the Highlands was a deep canyon filled with rubble. Berkey, looking at the core borings was the first to conclude that glacial scouring, not stream erosion, had created the channel through the Highlands.

This was an important finding. Ice valleys erode in a different profile than stream valleys, affecting assumptions about the shape of the river bed. Stream valleys have a deep notch in the bottom, whereas ice valleys are generally U-shaped and flat-bottomed. If the gorge was an ice valley, its depth could be determined as soon as one part of its floor was touched. If the gorge was a stream valley, the slopes of the two sides would help determine the slope of the bedrock floor.

Armed with the information from the first sets of borings, Berkey suggested another method to determine the depth to bedrock. While the scows continued their channel borings, test shafts were sunk on each bank of the river, and, at a depth of 250 feet, inclined bore holes were drilled out from each shaft toward the center of the river.

This method proved no less problematic than drilling through the river channel. The first contract called for a set of holes to pass at an elevation of −1200 feet. The hole from the west shaft started at an angle of 38 degrees, while the one from the east shaft was set at 43 degrees. In those days, before the advent of sophisticated calibration equipment, vials of hydrochloric acid were used to obtain the inclination of the holes. After every hundred feet of drilling, the vials were inserted into the bore hole. The acid etched a line in the glass tubes showing the angle of the hole, so that any necessary corrections could be made. Drilling deep into bedrock created other problems. It caused the loss of time, equipment, and core borings. During this process, two of the diamond drill bits overheated. One was recovered, but had become so hot that the steel and diamonds had fused to the rock. The other, valued at $1500, was sunk to a depth of 1834 feet and could not be not recovered, although crews spent two months trying.

However, the results of this new method provided critical information. The first set of cores passed through solid rock, proving that the river bottom did not extend below that depth and might even be considerably higher. A second contract was drawn up, calling for a new set of drilled holes to cross at an elevation of about 900 feet below sea level. This set of borings actually crossed at an elevation of 955 feet below sea level, and their cores once again showed continuous, sound granite.

The engineers concluded that the tunnel could safely be constructed 1100 feet below the surface, allowing at least 150 feet of sound rock clearance at the lowest point in the gorge, to resist the enormous bursting pressure of the water. All further attempts to determine the actual ceiling of the bedrock from scow-mounted drills were abandoned.

Construction of the tunnel proceeded immediately. The test shafts that had been excavated for geological exploration were enlarged for use as downtake and drainage shafts for the tunnel. Predictions of drought increased the urgency of completing the Hudson River tunnel, and the Breakneck shaft was sunk with record-breaking speed. Using new methods and equipment, a total depth of 588 feet, averaging 16.5 feet in diameter, was sunk in 93 working days at the rate of 6.5 feet per day, reaching tunnel grade in late 1910. The significance of this achievement is illustrated by comparison with the construction of the railroad in 1848, in which drilling progressed at the rate of one to two feet per day.

The tunnel below the river was excavated from both sides, with crews working toward the center. As they worked, they could often hear each other

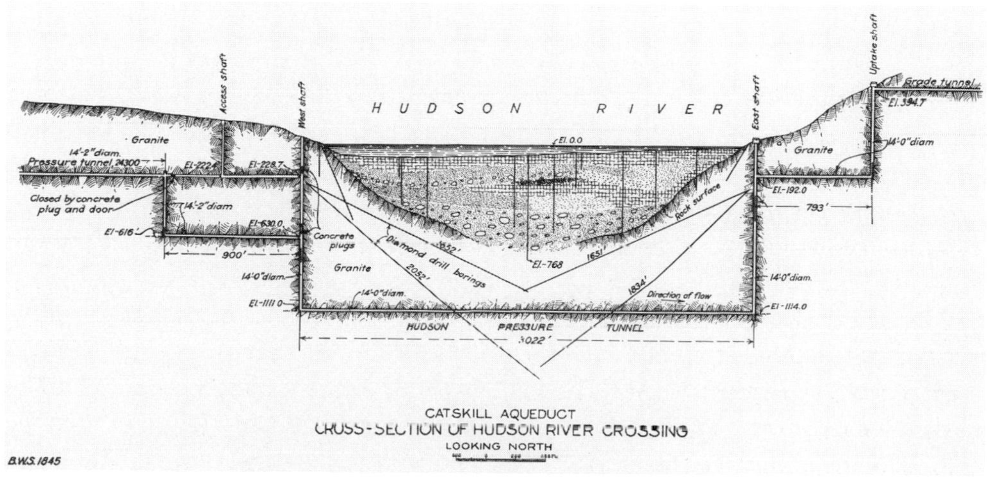

Cross section of the Hudson River crossing of the Catskill aqueduct looking upstream, showing the location of the tunnel and the diamond-drill borings made to explore the rock to determine its location. NYC Board of Water Supply. Courtesy of Yale University Library.

Drilling the Storm King tunnel of the aqueduct. New York City Board of Water Supply. Courtesy of Yale University Library.

approaching through hundreds of feet of rock. When they broke through the headings, their survey lines would often meet within fractions of inches.

Mining technology was used to clear the muck and rock debris from the tunnel excavations. Narrow-gauge rail tracks were laid in the tunnels, and mules were used to haul mine tipple cars from the heading of the tunnel to the shafts where the debris would be transferred to a hoist. Eventually, small electric locomotives replaced the mule trains.[6]

To assure the smooth, watertight lining necessary for the siphon to work, the tunnel was lined with concrete. Huge steel forms were assembled in the tunnels, and concrete was poured between the steel forms and the tunnel walls. It took 46 working days to line the Breakneck tunnel shaft.

Construction of the tunnel was not without its hazards. The explosion of a powder magazine in the Breakneck Valley, "levelled everything in the neighborhood and blew the fragments of the attendant over ten acres."[7] The cause of the accident was never determined. On the west bank, a condition called "popping rock," where internal geological stresses caused sizeable pieces of granite to pop off the tunnel wall, created several accidents and at least one fatality. Popping rock which was confined to the Storm King side, appeared only in greenish granites and never in the pink varieties or the diorite veins. Although apparently sound when first exposed to air, the rock later snapped off in flat, wedge-shaped layers. According to Lazarus White, "In both the Hudson and the Rondout tunnels the phenomenon gave great annoyance to engineers, who would find their scales and line plugs under a mass of rock fragments, after carefully selecting what they thought to be the most solid places." Eventually the popping rock was stabilized with steel plate. However, other problems continued to plague the project. When the Storm King tunnel was finally filled with water, the granite shifted, causing leaks. The entire works had to be pumped out and a new shaft drilled, bypassing the fractured zone. Some fifty years later, concern that blasting on Storm King would cause new leaks in the aqueduct lent support to the movement to preserve the mountain.

Work on the aqueduct was supervised by Division Engineers of the New York Board of Water Supply who each were responsible for a section of the route. Their work required that they move to the location of construction, and evidently several of them took advantage of their posts to enjoy the scenery. William Thompson Howell, who campaigned for the preservation of the Highlands in 1908, recorded in his diary for September 7–19, 1910 a camping trip to Cedar Pond in the Highlands.

> We had one visitor that day who had come to see us the night before—James F. Sanborn, Division Engineer with the New York Board of Water Supply, in charge of the construction of one of the northern sections of the great Catskill aqueduct, with a headquarters at New Paltz. He had come down to Monroe via the Wallkill Valley and Erie railroads, and walked in to our camp from Monroe. He turned off, after leaving Arden, on the wrong wood road, and finding himself in the valley west of our ponds, travelled across the pathless hills by compass. He came into camp Saturday afternoon, as we were preparing lunch, with a pack on his back in which were four pounds of sirloin steak and a bottle of whiskey.
>
> He had not been in camp ten minutes before he was insisting on being a member of a party which was going on a tramp of five miles and back in order to bring in some thirty quarts of oats for the horses expected on the morrow. . . .
>
> "A man of culture," said Baird of him, in his camp talk later, "and a Harvard graduate." Mr. Sanborn was also par excellence the camper and out-of-door man, and made me realize more fully than I had ever done before the fraternity of feeling which exists among men who go often into the open.
>
> We had looked for other visitors, among them William E. Swift, Division Engineer in charge of the great aqueduct tunnel work under the Hudson River at Cornwall, who had promised to bring Mrs. Swift and spend a night in camp. They were just preparing to start out, to drive across the mountains to our camp, when the unexpected arrival of visitors spoiled their plans, to our great regret."[8]

On January 30, 1912, New York Mayor Gaynor fired the shot which broke down the rock wall separating the east and west tunnels, and those present walked through a tunnel of solid rock 1100 feet below the surface of the river. Except for a few streaks of water here and there, the tunnel was bone dry. Eighty percent of the construction of the Ashokan system was complete.

On September 9, 1913, the reservoir dam gates were closed and the Ashokan's two basins began filling up. By the end of 1917, 250 million gallons per day flowed through the tunnel from Ashokan to New York. In 1924, the Schoharie reservoir was tapped, raising the level of flow to 500 mgd.

In 1917, the jubilant mayor appointed a committee of five hundred citizens to arrange for a celebration of the completion of first phase of the aqueduct system. The Catskill Aqueduct Celebration Committee had a commemorative medal struck with the inscription:

AN ACHIEVEMENT OF CIVIC SPIRIT
SCIENTIFIC GENIUS AND
FAITHFUL LABOR

A series of events were planned, including an allegorical pageant with dancers draped in flowing costumes to represent clouds, rain gods, and fire spirits delivering a symbolic gift of water from the mountains to the city. The celebration was cancelled because of World War I, but the New York Public Library produced an exhibition and a catalog which described the pageant in detail as it was planned to occur.

The construction of the aqueduct was a monumental achievement, and the problems conquered in crossing the Highlands gorge led to improvements in the speed and economy of sinking circular shafts; the use of concrete tunnel lining instead of timber; improvements in the method, economy, and speed of lining tunnels with concrete; the use of hammer drills; the employment of steel forms on movable carriages; and the use, for crossing valleys with water supply systems, of deep, pressurized tunnels instead of steel pipes. Although the construction of the Storm King/Breakneck tunnel left virtually no imprint on

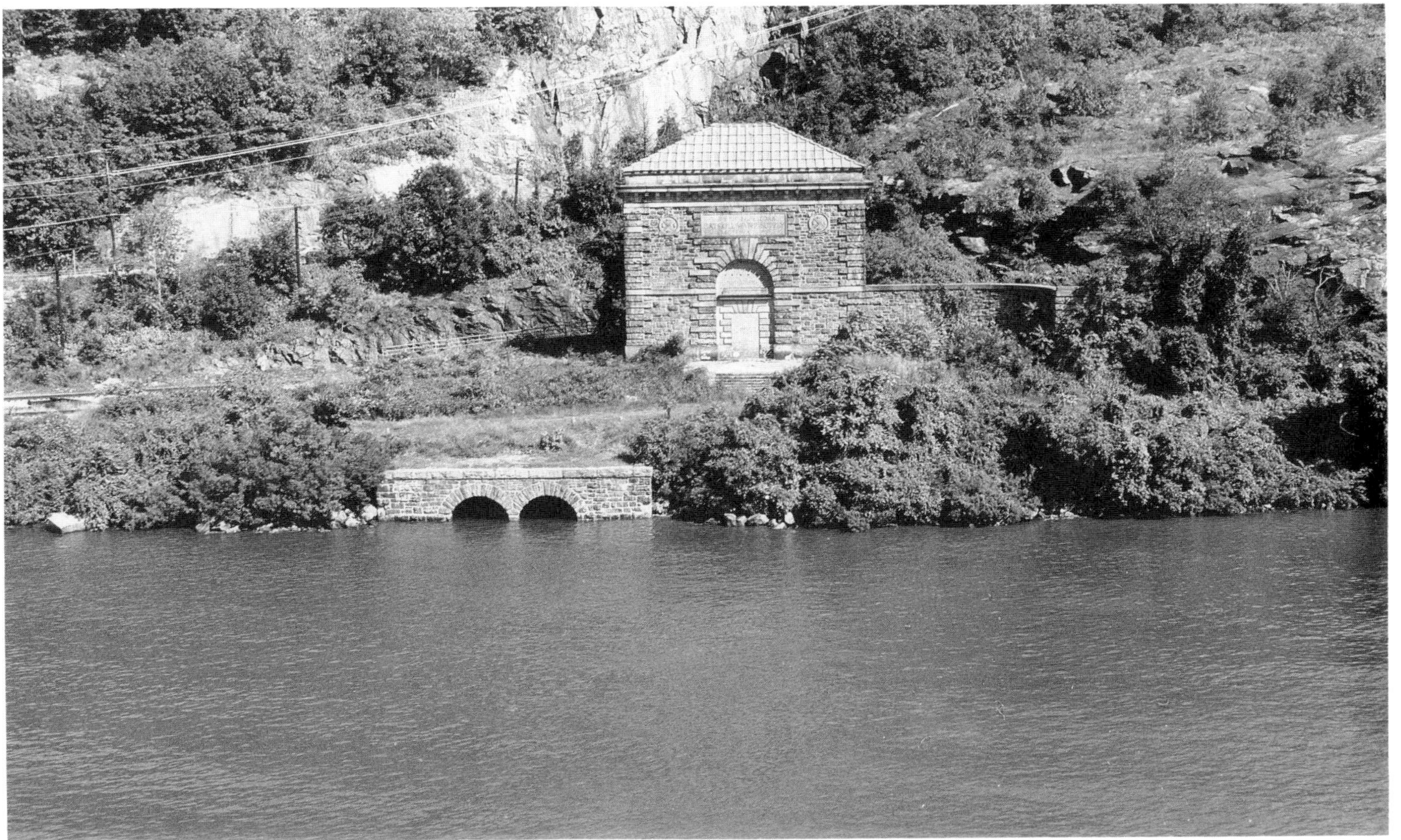

Stone structure at Breakneck Mountain, part of New York City's Catskill aqueduct system. This building, often thought to be a pump station, protects a large manhole cover and is the only remaining visible evidence of the aqueduct tunnel at the Storm King Breakneck tunnel. No pumps were used on the aqueduct. Copyright © 1990 by Robert Beckhard.

the Highlands' scenery, it had revealed the subsurface geology of the Highlands gorge, and this set the stage for the conservation of the Highlands gorge six decades later.

The other two engineering projects—the Storm King Highway and the Bear Mountain Bridge—changed the Highlands forever. As highway projects, they broke the region's isolation. Each highly visible, they left their mark on the mountain landscape. Although there was considerable enthusiasm for technological achievements at the time, both highway and bridge aroused a public outcry over the impact on the beauty of the Highlands. However, proponents argued that bridges and highways would enhance the enjoyment of scenery and, in the end, construction proceeded.

The first of these two projects was the Storm King Highway. The concept of a road along Storm King's rugged face was discussed as early as 1903, before Bear Mountain Park was created, and was stimulated by local concern, since the only way to get from Cornwall to West Point and Highland Falls was up and over the mountain on a long and tortuous road called the "S's." A short cut around Storm King's face would reduce the trip from about eight miles to four and would provide a link in the developing state highway system running north-south between New York and Albany. The proposed route was surveyed, but numerous engineering obstacles as well as political and bureaucratic ones delayed construction for years. It was not until the jurisdiction of the Palisades Interstate Park Commission was extended into the Highlands, and the Commissioners took up the cause, that the combined obstacles of nature, government, and politics were overcome. By comparison with the aqueduct tunnel, construction of the Storm King road was a minor undertaking, but for a highway project it was major. In terms of cost per mile at its time, the four-mile route was the most expensive in New York State history and took more than three years to build.

The most significant problem was the difficulty of working on Storm King's towering slope. The highway, as planned, would be no more than a narrow ledge benched out of the mountain's sheer cliffs overhanging the river about 400 feet above the water at its highest point. To survey the route in 1903 required a great deal of ingenuity. In order to plot the points, workmen had to use rockets which fired jars of paint at some of the most inaccessible areas. In other places, surveyors were lowered on ropes. Later, when construction began, it was found that much of the Storm King Highway could be built only by manual labor. Although sophisticated machinery was available by that time, cranes and other heavy equipment could not be brought onto the steep mountain face. Smaller machines had to be hauled up piecemeal on mule back and assembled on location. To blast the rock, it was often necessary to suspend the workmen in baskets held by ropes, while they drilled ten-foot holes for dynamite into the rock face. Blasting had to be timed to prevent falling rock from crushing passenger trains which used the track below. For most of its distance, the Storm King road is a notch carved into the mountain face. However in some locations, to maintain proper slope and drainage, the roadway had to extend out over the ledge on retaining walls. Constructed stone by stone by skilled workmen, these walls were built up to fifty feet high in some places.[9]

Nature was not the only obstacle in the construction of the Storm King Highway. First surveyed in 1903, work did not begin until more than a decade later, delayed by a taxpayer suit, opposition from the West Point brass, and eventually the advent of war. The decision to proceed with the route would almost certainly have been stalled indefinitely without the intervention of the Palisades Interstate Park Commissioners.

Most significant of these difficulties was the reluctance of the United States government to provide a right-of-way through West Point, connecting the proposed highway with Route 9W to the south. Successive West Point Superintendents steadfastly opposed the construction of the road, citing the potential for interference with artillery practice and questions of jurisdiction over the road. West Point preferred its isolation and wanted no part in the shared use of its facilities or road system. Repeated visits by Commissioners of the PIPC, arguing the benefits of park development on the future of the state, were rebuffed.

A further complication was the cost of purchasing the land needed for development of the road. The Storm King Stone Company operated a quarry on lands along the proposed route. Incorporated in 1906, three years after the road was first proposed, it appears that quarrying was only one of the motives for buying the land. In their offering to the stockholders, company executives pointed out that if the state condemned the land for highway purposes, the revenue would far exceed the original investment. They knew that a quarry on the site would only increase the government's urgency to purchase the land from them, and they intended to hold out for top dollar.

Concern over the costs of condemnation aroused considerable public discussion as did the question of scarring the venerable face of Storm King Mountain. In 1907 a taxpayer suit brought the condemnation proceedings to a halt, and the project was abandoned.

Finally public pressure on the state to construct a north-south route could no longer be ignored, and in 1914 plans for the present highway were drawn. This time, with active lobbying by the Palisades Interstate Park Commission, a deal was struck between the state and federal governments over the route from Storm King through West Point to Bear Mountain, clearing one of the many obstacles to construction of the road. One of the conditions of the deal was that the road be closed for two hours a day for artillery practice. Another was that the federal government would retain jurisdiction of the road where it passes through West Point.

The problem of condemnation cost was resolved when two leading families of Cornwall, the Tafts and the Stillmans, threw their support behind the project and donated the rights-of-way over the lands which they owned. Taft, a major developer in the area and owner of a home construction company, embarked on a publicity campaign, writing dozens of articles in magazines and newspapers, stressing the importance of the road in bringing motorists to the river's edge.

While Taft stood to benefit from improved highway access, Stillman did not; however, he joined many of his peers in supporting the road as a way of promoting park development at Bear Mountain. Shortly after the road was completed, Stillman's son Ernest gave to the PIPC 800 acres of riverfront land bordering five miles of the highway, creating a new Storm King section of the park. At the condemnation hearings for the quarry land, PIPC stepped in and offered to pay one-half the costs, derailing the opposition of the concerned taxpayers.

To those who argued that the view of Storm King would be marred, the Park Commissioners countered that the highway would allow thousands of motorists to view the Highlands from Storm King's heights, a greater benefit. Shortly before the highway was opened in 1922, Dr. Edward Partridge, a member of the Palisades Interstate Park Commission, wrote that motorists would soon enjoy a scenic highway "of spectacular interest." He noted that the Albany Post Road on the east bank was set too far back in the hills to afford comparable scenery. "As far as the view is concerned the traveler might be a hundred miles from the river."[10] The Storm King Highway would solve that problem. As the most powerful protectors of scenery in government, the Park Commissioners' views prevailed.

The Storm King Highway. Copyright © 1990 by Robert Beckhard.

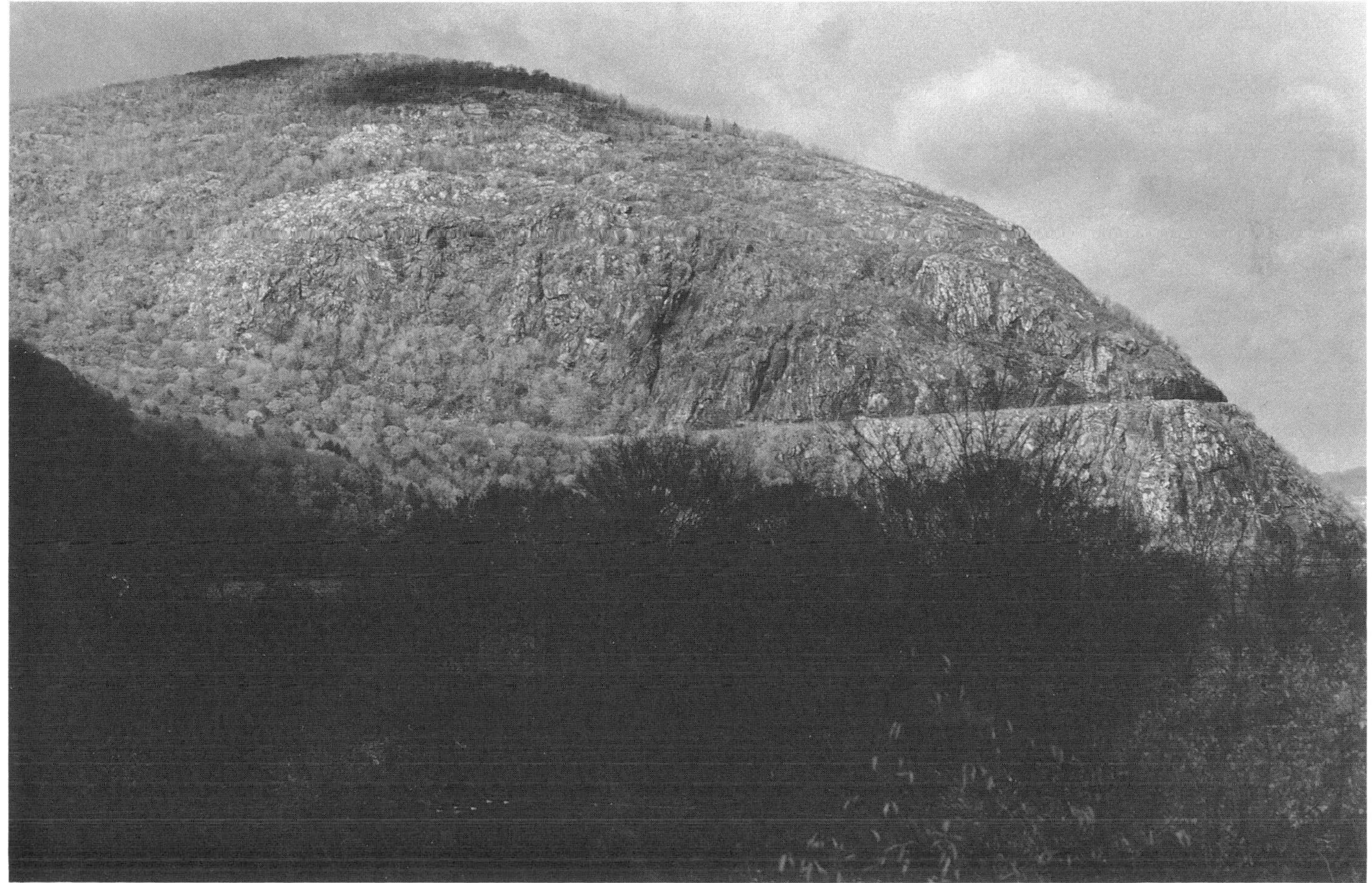

The plans for the road were finally approved in 1914, but the advent of war delayed construction until 1919. Three years later, in 1922, the road was opened to traffic and eventually became a link in the north-south through road between Albany and New York.

The Storm King Highway is indeed one of the the most scenic roads in the Hudson Valley, but it scars the beauty of the northern gateway to the Highlands. A major feat of engineering in its time, it enjoyed a brief history as a through road until the opening of the "Storm King bypass" (Route 9W) in 1940 made it obsolete. It is now used primarily as a local route between West Point and Cornwall and as a scenic route for those who seek to enjoy the Highlands from an elevation of 400 feet.

Like the Storm King highway, the Bear Mountain Bridge was a response to the growing traffic problems created by the newly popular and affordable motor car. A highway system developed rapidly on both shores of the Hudson, but the only bridge for cars crossing the River was north of Albany. South of Albany, there was only one Hudson River span, a railroad bridge at Poughkeepsie, opened in 1888.

For New Yorkers traveling to the park, this meant a long wait at ferry

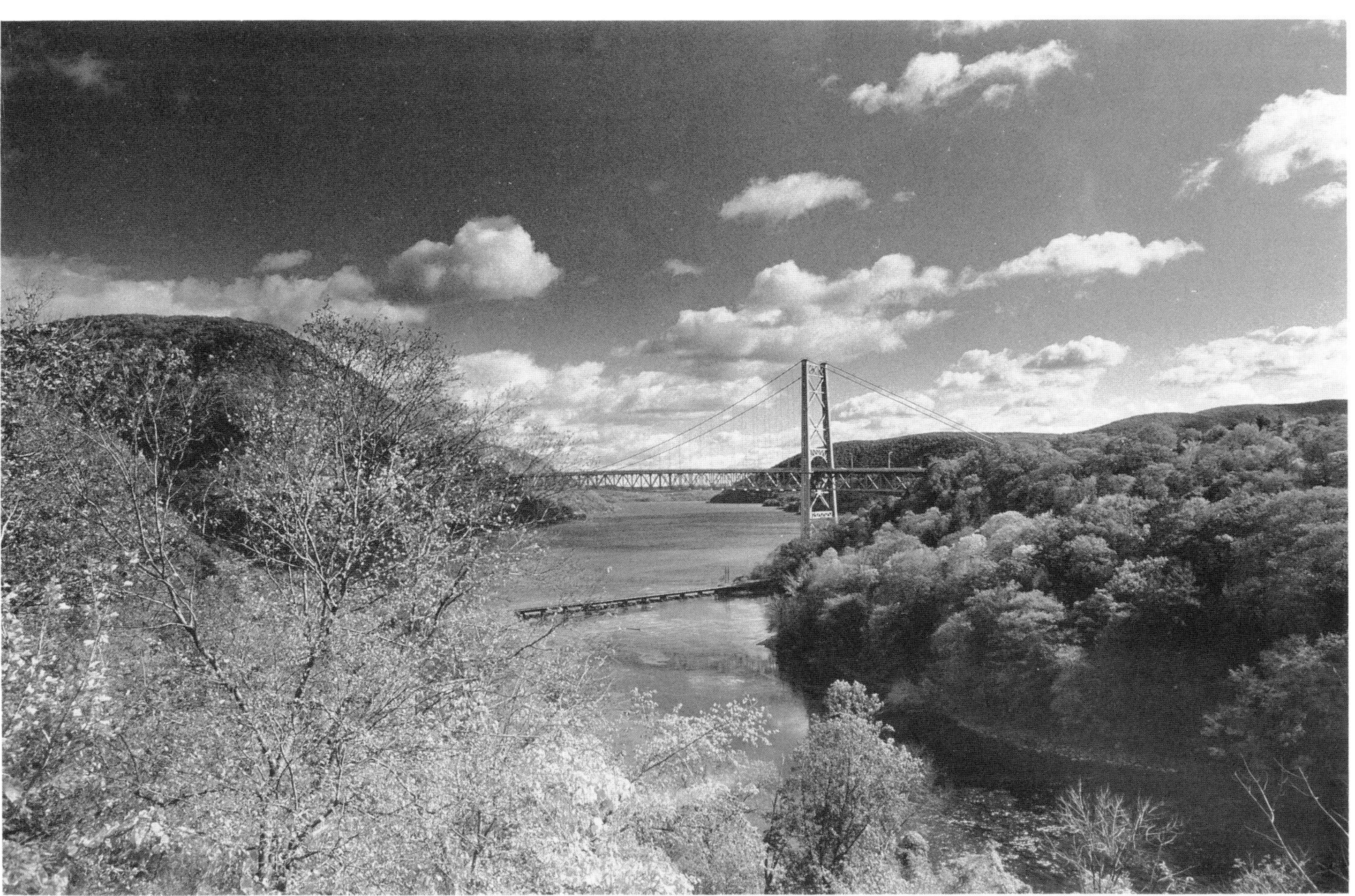

The Bear Mountain Bridge. Copyright © 1990 by Robert Beckhard.

landings. A February 10, 1922 article in the *New York Times* observed that ferry service was woefully inadequate resulting in long lines of cars waiting far into the night. "Even the hotels in the river towns cannot take care of the stranded ones when there is a glut of traffic. People have been known to sleep out in their cars all night." As the pressure for a river crossing intensified, the technology for building suspension bridges was advancing to the point that a short span across the river could be attempted. The Highlands, where the Hudson was narrowest, were the most likely location and the need to provide access to Bear Mountain Park would be justification. In February 1922, a bill was introduced in the state legislature which authorized the creation of a corporation to build a bridge across the Hudson in the Highlands. The corporation, a private entity financed by stockholders, was to construct the bridge and operate it for a 35-year period, after which the structure would become the property of the State. At its own option, the State could take over the bridge at any earlier time for a price specified in the law. In support of the bridge, it was argued that park usage at Bear Mountain would double, and that severe traffic congestion would be relieved. The bridge would also help, in the event of a railroad and ferry strike, to allow provisions to be brought into New York City, and it would help to stimulate growth in the surrounding countryside. The cost was estimated at $10 million, although when completed it actually cost about half that much.

To recover its investment, the legislation allowed the company to collect tolls for persons, vehicles, and animals. It cost 20 cents to transport horses and cattle, while calves, hogs, sheep and lambs could cross for only 10 cents. People on foot and passengers in cars were charged 15 cents, children 10 cents. Cars were assessed 80 cents, motorcycles with side cars 35 cents. The rates for horse-drawn wagons ranged from 60–75 cents depending on length.

The bridge was to be constructed where the river narrows between Anthony's Nose and Bear Mountain, the exact location where colonial forces had stretched the first of two chains across the river. The west end of the bridge would be adjacent to Fort Clinton, the colonial redoubt destroyed by British forces in 1778. Because of its proximity to Revolutionary War historic sites, some suggested that the span be called the George Washington Bridge.

Extending 1600 feet across the Hudson, the bridge would be of unprecedented length. The narrow water gap in the Highlands was the only place on the river south of Albany where a suspension bridge for automobiles could be tried. The attempt was made possible by rapid advances in the engineering of this type of bridge.

The first suspension bridges constructed in this country date from as early as 1796. They were built of wood and iron, using wrought iron chains as suspension cables.[11] It was not until 1844 that the modern method of using wire suspension cables was introduced by John A. Roebling for a bridge used to carry water and boats on the Pennsylvania Canal across the Allegheny River at Pittsburgh. It consisted of seven spans, each 162 feet in length, supported by parallel wire cables seven inches in diameter. In following years, the Roebling company developed the technology for successively longer spans and greater capacities, and introduced the French method of spinning cables in place, used in 1854 for a bridge over the Niagara Falls rapids—an 821-foot span supported by four ten-inch cables. The combination of a railroad and highway suspension bridge was a feat, said the Roebling company, "then thought impossible."[12] The string which formed the first connection between the two banks was flown across the gorge by kite.

A year later, in 1855, the Bessemer process for making steel was invented, and the improved Siemans Martin process soon after that. Steel, however, did not replace wrought iron in bridge construction until the turn of the century. The Eads Bridge at St. Louis, built in 1874, was the first to employ structural steel. This structure together with the Brooklyn Bridge, built in 1883, issued in the use of pneumatic caissons. This technological advance allowed bridge foundations to be sunk to greater and greater depths. The Brooklyn Bridge also pioneered the use of steel as a material for bridge wire and the use of galvanized coating for bridge cable wire, utilizing 14,568 miles of wire cable to support a main span of 1595.5 feet and a total span of 3,455.5 feet, including side spans.

Methods of spinning cable wires were constantly being improved, reducing the construction time dramatically. The Brooklyn Bridge, which used 900 tons of wire per cable took 21 months to produce. Twenty years later, in 1903, the cable wires of the Williamsburg Bridge were produced in seven months, while the cable wires of the Manhattan Bridge (1909), employing 1600 tons of wire per cable, took only four months to produce.

It was advances such as these which inspired a group of entrepreneurs to consider construction of a bridge across the Hudson in the Highlands. The construction of the Bear Mountain Bridge was financed privately. Its principal backer was a scion of the same family whose philanthropy had created the park: E. Roland Harriman, son of E. H. Harriman and brother of Averill, who would later be Governor. Harriman established the Bear Mountain Hudson River Bridge Corporation in 1923 and became its president. The company raised more than $5 million in capital to pay for construction, to be repaid by collecting tolls.

While the purpose of the bridge was to make the park more accessible to motorists, it had every prospect of being a successful business venture. The number of cars at the park had jumped from two thousand in 1916, when the park opened, to over a million by 1921, and with traffic jams plaguing every ferry crossing, the potential revenue was enormous. It was estimated that the average family of four visiting the park would have to pay $1.15 in tolls to cross the bridge. Newspapers reported that the bridge was constructed to handle five thousand automobiles per hour.

It was not long before Harriman's motives and the relationship of PIPC commissioners to the project were questioned. The cause for this stir was the design for the bridge. It was criticized by Lawrence Abbott as an unsightly structure which would mar the scenery of the Highlands. His views were published in a July 1923 issue of *The Outlook*, which printed a photograph of the proposed design. Abbott wrote "With all due respect, it looks, in its wonderful setting of mountains and river, like a piece of tin frumpery." He also noted the rumor that several Commissioners of the Palisades Interstate Park had joined Harriman as investors and suggested that if this were so, they might be more concerned about keeping their costs down than protecting the public interest in the scenery.

Following Abbott's article, the *New York Times*, in a July 20, 1923 editorial titled "An Infliction of Ugliness on the Hudson," blasted the Bear Mountain Hudson River Bridge Company for its plans to construct "a bridge wholly out of accord with the scenery around it and indicative only of a desire on the part of the builders to make it as cheaply as they can." The *Times* observed that "Ugly bridges, indeed, have been built in modern and ancient times, but they are so few that apparently it is easier to make them beautiful than ugly, and that to attain ugliness . . . no small amount of ingenuity and determination must be misapplied.

"There is no excuse at all, consequently, for what is about to happen near one of the most admirable manifestations of natural beauty in the vicinity of this city, and in this part of the country, for that matter—the erection of a bridge across the Hudson that will be an offense to the eye for as long as it stands."

In a *Times* article the following day, J. B. White, President of the Palisades Park Commission, defended the design of the bridge, noting that the Commission had also questioned the design and had "engaged the best architectural advice we could get," and had required design changes before approving it. He contended that the picture Mr. Abbott had showed was out of date and did not represent the plan finally adopted by the Commission. He called the reports of conflict of interest "contemptible," suggesting that PIPC Commissioner Averill Harriman "should get a Congressional Medal for putting money into this bridge." On July 24, E. Roland Harriman, President of the Company, ended the debate with the following statement which was printed in the *Times*: "It is felt by those responsible for the affairs of the bridge company that they are willing to rest their case on the merits of the design when the project is complete."

Construction proceeded, and as Roland Harriman predicted, the completed bridge was indeed exquisitely designed. By today's standards, the size, scale, and location seem a perfect reflection of its intended use. A narrow, two-lane bridge, held aloft by a graceful sweep of cabled steel, and complemented by its Victorian toll house—a cottage built of stone with a steep slate roof, diamond-paned windows and a tower—it serves no interstate highway. It is, instead, a grand entry to a grand park.

The Bear Mountain Bridge, when it opened in 1924, was believed to be the longest suspension span in the world, exceeding the main span of the Brooklyn Bridge by 400 feet; but it held that record only briefly. Its 1600-foot span was surpassed two years later by the Philadelphia Camden Bridge at 1750 feet, and the George Washington Bridge, built on the Hudson only eight years later, was nearly 3,500 feet in length.[13]

The cable wires and steel ropes for the Bear Mountain Bridge were manufactured by John A. Roebling's Sons Company which published in 1925 a

Construction of the Bear Mountain bridge suspension cables photographed by the Roebling company. Courtesy of the Palisades Interstate Park Commission, Bear Mountain.

pictorial account of the step-by-step construction of the span in a booklet entitled "Construction of Parallel Wire Cables for Suspension Bridges," which notes that in addition to being the longest suspension bridge known, the Bear Mountain pioneered new engineering principles in cable construction:

"The construction of cables for the Bear Mountain Bridge may be said to have been the turning point in history of parallel wire cable fabrication, as several new theories, put to a rigid test during the construction of these cables, proved very successful, and opened the field to cables of practically any diameter and number of wires." The principal advance, according to Roebling, was the use of flat wire bands, which eliminated a slow and tedious compacting procedure in the construction of cable wires. With these improved methods, the Bear Mountain Bridge set yet another record, with its wires being strung in three and one quarter months—quadruple the rate of stringing on the Brooklyn Bridge and double the rate on the Williamsburg Bridge.[14]

The bridge was designed by Howard C. Baird. William Finch Smith was the Chief Engineer and it was constructed by the firm Terry and Tench. There were 125 riveters working on the bridge, which at high tide offers a clearance of 155 feet for passing ships. The dangerous work of constructing the span was accomplished without the loss of a single life.

On September 10, 1924, the first car crossed the bridge, driven by M. Belknap, the engineer in charge of construction. The official opening occurred two months later, on November 23, 1924. For the opening ceremony, 400 automobiles assembled. Accompanied by the West Point military band, they paraded from Peekskill, along the Anthony's Nose Highway across the bridge to a bronze tablet which was unveiled by Mary Harriman. In triumphant celebration of the occasion, the military band at nearby West Point played the "Star

Spangled Banner." The enormous difficulties of spanning the Hudson had been conquered.

The celebration was a public relations *tour de force.* After an incredible media build-up, the bridge was opened to the public the following day for Thanksgiving. For two dollars the Bear Mountain Inn offered a traditional Thanksgiving turkey dinner, accompanied by music. On its first day, five-thousand motor cars crossed the new bridge.

Once completed the Bear Mountain Bridge opened up connections with the Storm King Highway on the west side of the Hudson and created another scenic drive to the east, a new three-mile section of highway which was benched out of the rock along Anthony's Nose, connecting the span with Peekskill. At roughly the same elevation as the Storm King Highway, the Anthony's Nose Highway, now Route 6, with its scenic overlook at Inspiration Point, was deemed to be a "worthy rival" according to press accounts of the day. This highway completed a north-south section of highways designed to allow the motoring public to enjoy the spectacular scenery of the Highlands throughout its entire fifteen-mile length.

Although it achieved its public purpose, the Bear Mountain Bridge was nevertheless a financial disaster. It operated at a loss for thirteen of sixteen years until it was sold to the state in 1940 for about $2.3 million. In recommending to Governor Lehman that the state purchase the bridge, Robert Moses, Chairman of the State Council of Parks, stated that "It is unnecessary to emphasize the interest of the State Park Council in this bridge which is one of the main approaches to Bear Mountain Park and to the entire Palisades Interstate Park system. . . . No good purpose would be served by a complete history of the financing, building and operation of this bridge nor by analyzing the motives of the group responsible for it. The fact must be stated, however, that there is nothing in the financial set-up of the Bear Mountain Bridge to show that this was, at any time, a genuinely philanthropic or altruistic enterprise."[15]

The highway system which led to the park had a far-reaching effect on the Highlands, opening the region up to development. But this was not a concern of those who built it. Interviewed for the Associated Press on the fiftieth anniversary of the bridge, Frank Manthey, Bridge Manager for forty-seven years, recalled that "It was the first highway bridge over the Hudson below Albany, and it meant that for the first time in the 150 years of the Republic, you could cross the river without waiting for a ferry—which didn't even run when the Hudson was frozen over.

"What we learned here, others came to see—and it helped them to build the Holland Tunnel, the Lincoln Tunnel, the George Washington Bridge and every other crossing up and down the river ever since."[16]

11. Dynamite, Depression, and Design with Nature

THE CREATION of Bear Mountain Park and the construction of the Storm King Highway brought quarrying to a halt on the west shore of the Highlands, and after the construction of the Storm King tunnel of the Catskill aqueduct, concern that blasting might endanger a vital link in the city's water system had prevented further quarrying at Breakneck on the east shore. However, none of this had decreased the demand for stone, and after World War I the trap rock companies began to cast about for new opportunities. On the east bank of the Highlands, except at Breakneck, there were numerous spots to meet their needs, and in 1931 the Hudson River Stone Corporation bought 1000 acres on Mt. Taurus (Bull Hill), extending from the river to the summit, and began quarrying at a speed never previously witnessed.

The actions of the Hudson River Stone Corporation created an immediate public outcry, and after a twenty-year lull, feverish efforts to conserve the Highlands began again. In the 1930s, governors and legislators would again come to the region, this time not for ribbon-cutting, but to develop new programs for the protection of the Valley's history and scenery. During these years, which coincided with the Great Depression, far-sighted concepts for preserving the Hudson were developed, and many of them remain as models for conservation today. The advent of war forced these efforts to be abandoned. However, in this short ten year period, major advances were made, most of them through the actions of private citizens as well as relief workers drawn from the ranks of the unemployed. When peacetime returned, new issues in conservation, primarily water pollution, caught people's attention. Preserving the land became less pressing than cleaning fouled waters, but the image of the Highlands as a special place remained, and in the 1950s people once again worked their fantasies on the face of the land.

Public concern over the destruction of the Highlands' world-famous scenery by the quarrymen had been building for nearly a century before the acquisition of Mt. Taurus by the Hudson River Stone Company. One of the first Highlands' landmarks to go was the "Turks Face," a famous promontory on the ridge of Breakneck Mountain which took its name from its resemblance to a human face. One of the many points of interest for travelers, the Turk's Face was humorously described by John Maude in his 1800 journal of travels aboard a Hudson River sloop: "The profile of the Face Mountain so strongly resembles the profile of the human face, that I had for some time my doubts whether art had not assisted in improving the likeness. I have seen other blockheads which did not possess so sensible a countenance."[1]

In 1846, Captain Deering Ayers blew Turks Face to pieces. With a single blast, 10,000 tons of scenery crumbled into an undistinguished heap of rocks. The only consolation to the public was that the Captain was himself blown to bits a few years later while checking out an explosive which had failed to go off, an event that was noted with some satisfaction in guidebooks.

The introduction of dynamite in the 1870s increased the volume of activity considerably, threatening to reduce majestic mountains to rubble in a very short period of time. Nonetheless, public opposition to quarrying did not begin to solidify until the turn of the century. In 1894, an Englishman observed in a letter to the editor published in the *New York Times*: "I am astonished to find that the very scenery the Americans talk so much about is being defaced, desecrated and destroyed, and I cannot find any American who seems to care much about it."[2] Public concern began to surge within months of his commentary, but initial efforts to stop the stone crushers focused on the Palisades, and as the new Palisades Interstate Park Commission purchased more and more land, the quarries moved farther north, putting pressure on the Highlands, where the conditions for extracting trap rock could not have been better. The Highlands were close to markets that demanded ever larger amounts of rock year after year. The granites and gneisses of the river banks could produce a high-grade crushed stone for construction of buildings and roads with concrete, and the quarried rock could be easily and cheaply loaded for transportation by barge or train. As a concerned state senator at the time observed, "from the point of view of the quarryman the Highlands of the Hudson are absolutely ideal in every way."[3]

Quarrying on Breakneck had ceased in the 1850s, soon after Captain Ayers destroyed the Turk's Face, and was not resumed before the aqueduct put the mountain off-limits. However, Mt. Taurus remained unprotected, and its purchase by the Hudson River Stone Company in 1931, jolted the public consciousness.

1938 photograph of the quarry at Mount Taurus. Courtesy of the Hudson River Conservation Society.

The response by citizens and government was so quick that it almost seemed that they had awakened from a dream. Within months, Governor Franklin D. Roosevelt appointed a committee of the State Council of Parks, headed by Robert Moses, to study the possibility of raising funds through philanthropic gifts for state purchase of Mount Taurus as park land. Ironically, the previous owner of the property, Southard and Stern of Cold Spring, had offered to sell the land to the state for a park just a few years earlier, but the state had not responded.

The appointment of the Roosevelt/Moses committee seemed to assure that this time the state would act. The headline of the lead story in the May 1931 issue of *Scenic and Historic America*, the newsletter of the American Scenic and Historic Preservation Society, states "Mount Taurus Quarry to be Stopped—Protests Against Defacement of Beauty of Hudson Gorge Result in Action to Prevent or Transfer Blasting." But 1931 was two years after the stock market crash. The state was without funds for conservation, and the committee was unable to raise sufficient donations to acquire the land from the Hudson River Stone Company.

For the next five years, the blasting of dynamite continued to reverberate through the Highlands until a gash on the face of Mount Taurus dominated the view from West Point, the Storm King Highway, and the river below. On the

waterfront, vast piles of trap rock, former mountain scenery, awaited shipping to markets in New York and Albany.

In 1936, with a growing sense of frustration and outrage, twenty-nine residents of the Hudson Valley decided to do something about it. At a meeting in New York City they formed the Hudson River Conservation Society, electing William Church Osborn as president. "Organized purely by indignation at one piece of vandalism," they determined "to preserve the beautiful and historic Hudson Valley" to the best of their ability.[4] The certificate of incorporation set forth the following purposes:

> to preserve in its natural state the Hudson River and its valley and to preserve the historic landmarks thereof;
>
> to make a continuing survey of the Hudson River and its Valley as to conditions affecting health and recreation;
>
> to create public opinion and sentiment on all matters affecting the Hudson River and its valley, and to take public action thereon either directly or through community organizations.

The roster of Directors of the HRCS in 1936 shows that concern for the preservation of the Highlands was being passed on from the nineteenth century titans of industry to the next generation. It lists such names as Mr. and Mrs.

Samuel Sloan, Mrs. Charles Tracy, Mr. and Mrs. William Church Osborn, and Mr. and Mrs. George Perkins. In addition, the ranks of the HRCS membership, which soon numbered in the hundreds, extended to include Henry C. de Rham, Mrs. Herbert Frazier and Col. S. Benjamin Arnold of Garrison, Mrs. Cleveland Dodge (a relative of Osborn), and the Ledouxes and Stillmans from Cornwall. Other prominent families outside the Highlands also joined the Conservation Society, including the Aldriches and Delafields from Red Hook, the Hasbroucks from Kingston, Miss Margaret Reese from Hughsonville, Miss Faith Rockefeller from Greenwich Connecticut, and Alfred Olcott of New York, owner of the Hudson River Day Line. Many of these families remain active in Hudson River conservation today.

Being men and women of influence, the members of HRCS immediately took their cause to the highest levels of government, and their voices were heard in the halls of power. In November 1936, president William Church Osborn and vice-president General John Ross Delafield approached Governor Lehman in Albany. As a result of the meeting, Lehman proposed to establish a state commission to study ways of protecting the Hudson aimed not only at ending the quarrying on Mt. Taurus but also addressing the need to protect the historic resources and scenery of the river valley as a whole. The Governor's Annual Message to the Legislature in January 1937 announced his proposals with the following statement:

> The Legislature should give thought to a comprehensive plan for the preservation of the natural beauty of the Hudson River and for its improvement and development. The Hudson River and the Hudson River Valley are among the great scenic, historic and commercial assets not only of the state of New York but of the country as a whole. While parts of the River have been permanently protected against destruction or injury, the State has no control over the scenic and historic treasures of the greater part of the River.
>
> "Within the last few years we have witnessed the operation of a commercial quarry that is marring the beauty of an important part of the historic highlands. The commission which I propose should be charged with the duty of investigating the advisability of acquiring this quarry, protecting the scenic beauties of the River and the Valley generally and the possibilities of the further commercial development of the municipalities bordering on the River.[5]

Following the Governor's lead and with a now-active constituency group, Hudson Valley senators and assemblymen formed a caucus and began defining the problems confronting the region. With their backing, two bills were passed. One, introduced by Senator Erastus Corning, 2d, created a temporary Hudson Valley Survey Commission, and the other, introduced by Assemblyman Lee Mailler and known as the Mailler Law, authorized the state to accept gifts of scenic properties in the Hudson River Valley. (Senator Corning was 27 years old at the time; he is better remembered for his tenure as Mayor of Albany from 1941 until his death in 1983.)

The Valley Commission law established a nine-member commission and charged it with making a comprehensive study of the scenic and historic sites of the Hudson River Valley and recommending action to protect them from destruction or defacement. The recommendations were to include possible state acquisition of important sites. The Legislature appropriated $15,000 for the expenses of the Commission and the compensation of staff. It specifically required a study of preservation opportunities for Mt. Taurus and also authorized the Commission to "take an option on any privately owned site the acquisition of which it intends to recommend to the governor."

The Conservation Society invited Assemblyman Mailler and Senator Corning to speak about the legislation at the first annual meeting held at the Hotel Thayer at West Point on June 26, 1937. In his speech Senator Corning advocated a number of new ideas for conservation of the valley and suggested four major initiatives:

1. State purchase of Mount Taurus;
2. Development of a comprehensive plan for the whole river;
3. Private action to conserve land through deed restrictions;
4. State zoning for the river to prohibit defacement by quarrying through regulations which would permit quarrying only "beyond the skyline of the hills."

Corning stressed that "if the program of the State for the river is not a comprehensive one the purchase of Mount Taurus would not be of any great lasting benefit, because there are other places of great beauty in the Highlands that would be just as desirable from the standpoint of the quarryman. For real success there must be a plan to take in the whole river."

This was the first time that comprehensive planning for the river corridor had been advocated. It was also the first time that riverfront zoning by the state was discussed. The concept of a mile-wide protected scenic corridor in the Highlands had been proposed in 1909 in legislation supported by the Hudson-Fulton Commission, but that bill, which relied on state acquisition and ownership, never passed. Now, Corning proposed that the state's regulatory authority be used. The concept that the public's right to the protection of natural resources and scenic beauty was greater than that of private enterprise to their exploitation was being taken a step further.

Corning also discussed the idea of citizen action to preserve the scenery, noting that "private interests can do a lot in this work by putting restrictions in the deeds to any property disposed of by them. . . . Many large estates are now in the position that their owners are either unwilling or unable to keep them up. Till now the owner in such a situation had little he could do except sell the property for what it would bring and hope that the new owner would have the same patriotic and sympathetic idea of keeping our river beautiful that he did." The Mailler law, Corning noted, would make it possible for property owners to give their lands to the state "with the knowledge that they will always be kept in as beautiful and natural a state as possible." This, too, was a relatively new idea—that private groups could buy land, restrict its use through deeds or easements and then pass it on to the state to be managed as park land.

Corning, a pragmatic politician, saw the opportunities that preservation would create for economic development, especially in the real estate, recreation, and tourist industries. "Any real success in the plan to make the Hudson more beautiful will naturally bring back to its shores many of those who have left and others too who are looking for places of beauty in which to spend their vacations. With them also will come the over-night campers and picknickers. . . . Education is gradually teaching the people the value of conservation for themselves and for their children."[6]

Assemblyman Mailler also addressed the Annual Meeting of the HRCS and spoke specifically about the need to stop the quarrying in the Highlands. He pointed out that the state was spending hundreds of thousands of dollars to beautify the west shore of the Highlands but was allowing commercial interests to spoil the view on the opposite side. "Thousands of tourists from all over the world are annually attracted to the Storm King Highway from which, for several miles, the main view is that of Mount Taurus." His bill would allow public-spirited land-holders to donate a strip of riverfront property to the state which would permanently prevent its development without impairing the beauty of their estates.[7]

By appointment of Governor Lehman, Senator Corning was made chairman of the new temporary Hudson Valley Survey Commission. The Governor also appointed William Church Osborn of Garrison and Dr. LeRoy Kimball, the president of the American Scenic and Historic Preservation Society, as members.

As the Commission began its two-year study, the Conservation Society set about building its membership. By the end of the first year, 1937, there were 1130 members and $40,000 in the Treasury. When one considers that the HRCS was organized before the days of bulk mailings, this was a sizeable and well-financed group. To recruit members, the HRCS held garden parties up and down the river and set up local branches in each county and several riverfront towns. By 1940, the membership had swelled to an impressive 1860 donors.[8] Funds raised by HRCS were used to commission a study of potential quarry sites in need of protection. By this time, it was clear to members that state purchase of Mt. Taurus would not solve the problem. The stone companies would simply move to another site. In fact, reports of the American Scenic and Historic Preservation Society complain that the quarrymen deliberately purchased mountain properties on speculation that the state would buy them out.

HRCS began to identify and map potential quarry sites and to negotiate deed restrictions with the property owners. In its first year, the group protected 1200 acres in the Highlands—a 600 acre parcel owned by the Cornish family, which formed the northern border of the lands of the Hudson River Stone Company on Mt. Taurus, and another 600 acres owned by the Osborns in Garrison. Two years after HRCS was formed, the number of acres protected through deed restrictions on private property had climbed to 2350. In most cases the owners donated their mining and mineral rights to these lands or imposed the deed restrictions without compensation. However, not all owners of land were willing to do that, so in some cases, the Society raised money to purchase "danger spots" on the river which it then donated to New York State under the Mailler law. This included 250 acres of potential quarry sites on Breakneck and about 200 acres on Anthony's Nose, purchased with the assistance of the New York-New Jersey Trail Conference "so that the vital point of land, facing Bear Mountain Park and the Perkins Memorial Drive can never become another Mount Taurus." To protect the scenery and to provide recreation for the public, the Conservation Society also advocated the creation of a state park comprising the "Beacon Range" at the northern gate to the Highlands.

By 1938, with the protection of thousands of undeveloped acres secured, the problem of quarrying at Mt. Taurus still remained, and the members of the Conservation Society began to explore every means at their disposal to shut the operation down. Legal research revealed that permits from the state Land Board and the federal War Department were required for fill in underwater lands of the Hudson. In seven years of operation of the quarry, the Hudson River Stone Company had filled some twenty acres of underwater land with silt, rubble, and stone without a permit. The river fill had made it possible for the company to expand its operation to adjacent Little Stony Point. The Conservation Society hired an attorney, Judge Ellis Staley, who sued for an injunction on the operation of the quarry based on the illegal fill of the river. As a result of Staley's work, in 1939, the federal government required that by 1940 the company remove from the river all material down to the river bottom. The state government also fined the company and ordered it to refrain from dumping.

Meanwhile, the Hudson Valley Survey Commission was conducting public hearings throughout the valley and attempting to negotiate an option to buy Mt. Taurus from the Hudson River Stone Company. However in several areas of its inquiry, the Commission seemed to be drawing a blank, due to the strained state finances of the Depression era. Its 1938 *Preliminary Report* concluded that historic buildings could not be preserved because the state lacked sufficient funds to manage them. On the subject of quarrying, the Commission commended the work of citizen groups like the Conservation Society but concluded that as far as state action was concerned: "no recommendations can be made at this time that will stop the quarrying of the Highlands." The Commission concluded that conservation would come about only by the action of public spirited citizens and recommended educational efforts aimed at developing a conservation ethic among the people in the region.

In 1939, the Hudson River Survey Commission published its final report, which made no mention of quarries or Mt. Taurus but focused on promoting the public enjoyment of the river for recreation. By this time, a shift in public priorities was also taking place, and the primary concern of the legislature and a growing concern of the Conservation Society was pollution of the river. The commission suggested that relief workers assist the communities along the river with the construction of sewage disposal plants to clean up the pollution as soon as possible. The report contained a map showing sources of industrial waste such as milk products, bleaching and dyeing, powder laundries and paper wastes. It also recommended that Congress adopt legislation transferring au-

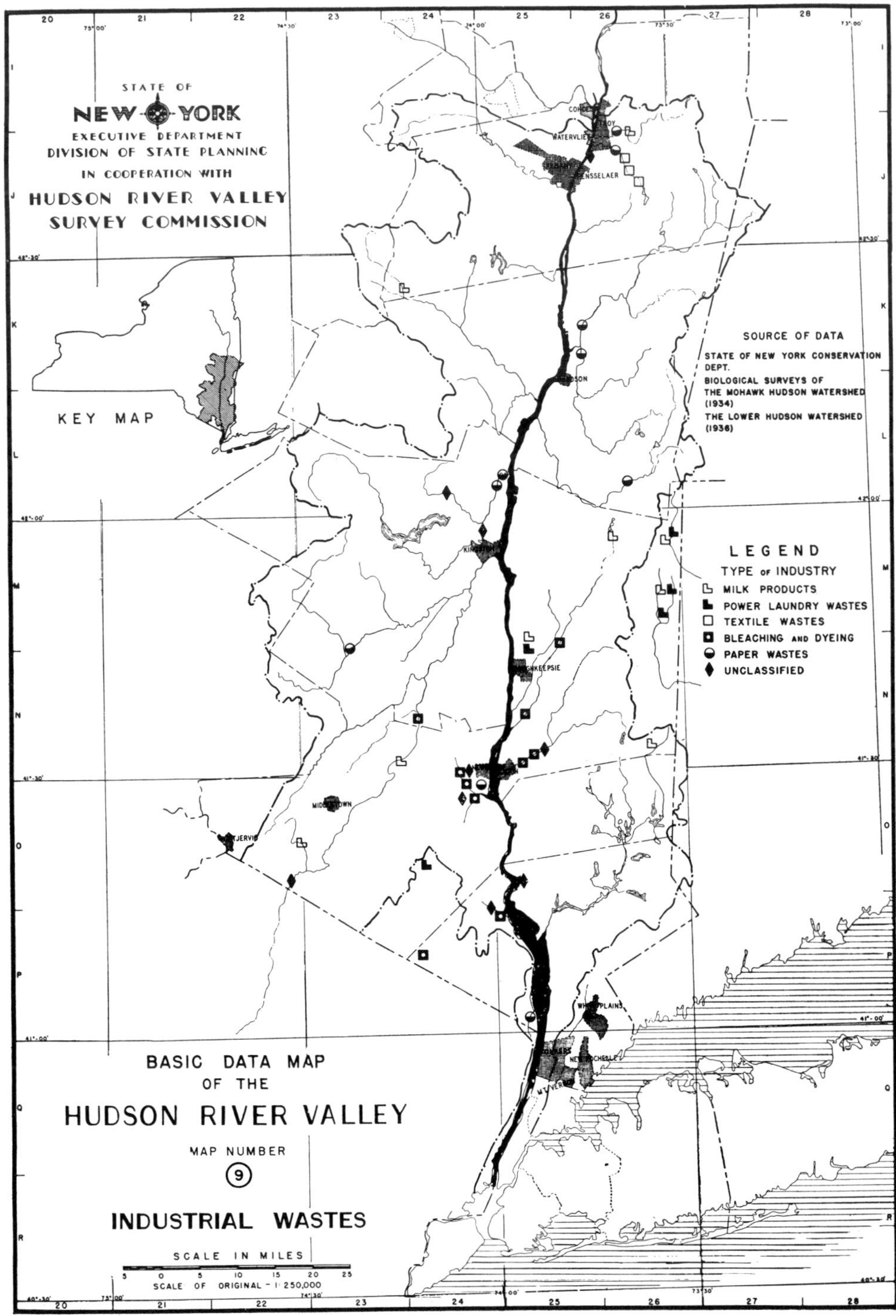

1939 map prepared by the Hudson Valley Survey Commission showing sources of industrial wastes polluting the Hudson including "milk products, power laundry wastes, textile wastes, bleaching and dyeing, paper wastes and unclassified" wastes. New York State Division of State Planning.

thority for the clean-up of pollution from the state to the federal government. Other recommendations included:

- studying the possiblity of building a scenic drive along the east bank of the river from New York to Albany "from which the superb natural beauty of the Hudson River could be made available to the people;" and

- increasing the number of marinas on the river to allow for greater recreation in light of the increasingly popular five-day work week and to promote economic development from sales of boats, outboard motors and supplies.

From the standpoint of scenic and historic preservation, the work of the Commmission was clearly a disappointment. The May 1940 issue of *Scenic and Historic America* , which was devoted in its entirety to Hudson River conservation and discussed at length the achievements of the decade, contained only two sentences mentioning the Commission report and findings. Describing it as a report "of value and interest" and "obligatory reading," the magazine discussed it no further.[9]

Like the Roosevelt/Moses Committee, the Hudson Valley Commission of 1937 was unable to stop the quarrying on Mt. Taurus. Its offer was rejected by the company as too low. However in 1944, during World War II, the mining operation shut down. The property remained in the possession of the Hudson River Stone Company until 1963, when it was purchased by the local electric utility.

While the study commission did little to live up to expectations, the federal government was quietly making a major impact on the conservation of the Highlands. Franklin D. Roosevelt, now President, had launched his massive public works program, and a vast army of the unemployed was mobilized to work on the construction of recreation facilities and trails at Bear Mountain State Park. The Civil Works Administration, and later the Works Progress Administration, provided relief workers who undertook the biggest construction program for the park since its creation.

During a five-year period of public employment programs, a system of pumphouses, reservoirs, sewer systems, vacation lodges, comfort stations, residences for park personnel, storage buildings, and the Bear Mountain administration building made their appearance. One of the most significant projects was the construction of a scenic drive to the top of Bear Mountain. Named the Perkins Memorial Drive, 95 percent of the work was done by hand labor.

Although new construction methods were available by the mid-thirties, this period of public works construction was characterized by faithful adherence to the naturalistic design principles which had been used at the turn of the century. The rustic architecture of the Bear Mountain Inn was echoed throughout the park in the solid structures of local stone, boulders, and timber constructed by the WPA.

Also during this time, some 3600 acres immediately southwesterly of Storm King Mountain were bequeathed to Harvard University by Dr. Ernest Stillman, who had specified in his will in 1940 that the land be used as a research forest. To assure that this purpose would be maintained, he provided a $1.5 million endowment in cash and securities. The land had been assembled by his father,

Construction of Perkins Memorial Drive in 1933 by relief workers. Courtesy of the Palisades Interstate Park Commission.

James Stillman, founder of what is today Citibank, for a colony of summer homes, and it was transferred to Harvard in 1949 when the younger Stillman died. Ernest, who had a passionate interest in silviculture, had numerous scientific papers printed about the natural history of the area. The *Black Rock Forest Papers* provide a valuable record of the changes in the land.

A great deal was done in the Depression era to support conservation. In the May 1940 Bulletin of the American Scenic and Historic Preservation Society, a special issue on the Hudson River, jointly published with HRCS, the two organizations offered the following thoughts characterizing the sentiment of the time: "Prior to the collapse of the post-war inflation in 1930, the prevailing public attitude was that while conservation was eminently wise and desirable, it was not a life and death matter. With the Depression, however, and the eagerness for national self-criticism it engendered, the condition of our natural resources was evaluated anew and another point of view accepted: that conservation could no longer be regarded as something that could wait until more pressing matters were disposed of, but was itself one of the most pressing and immediate necessities."[10] But even conservation is not as pressing as war. After a decade of zealous activity, the concern for natural resources and for the scenery of the Highlands was suddenly eclipsed in 1941.

During World War II, state and federal conservation efforts on the Hudson and in the Highlands came to a halt, and the Hudson River Conservation Society ceased further activities. In 1949, the Society reorganized and turned its attention to controlling water pollution, but it never regained its Depression-era vigor.

Another of the Highlands' more significant contributions of the 1930s was the number influential conservationists it engendered. History has particularly recorded the contribution of the Osborns, who as a family have been active in preserving the east bank of Highlands much as the Harrimans were on the west shore. Beginning in the mid-nineteenth century and continuing into the 1930s, members of the family bought up acreage—mostly small farms—in the Garrison area, that at one time totalled several thousand acres. William Church Osborn's brother, Professor Henry Fairfield Osborn, was for many years president of the American Museum of Natural History and was a moving force behind the pioneering Save-the-Redwoods League. His son Fairfield was for a generation the highly visible president of the New York Zoological Society and was instrumental in moving the Bronx Zoo to the forefront of the preservation of endangered species. In his day, he was known as one of the leading conservationists of the world; his book, *Our Plundered Planet*, was one of the first on the subject of worldwide conservation. His granddaughter, Robin Whyatt, served as executive director of Scenic Hudson for several years and until recently for the Natural Resources Defense Council.

William Church Osborn's dominant role in the establishment and early years of the Conservation Society and in the activities of the Commission have been discussed above. His daughter, Mrs. Vanderbilt Webb, succeeded him as president of the Hudson River Conservation Society. His son, William Henry Osborn II, succeeded Mrs. Webb in that role, and was one of the founders of Scenic Hudson. In 1974, William Henry donated a 656-acre property known as Sugarloaf Mountain to the Taconic State Park Commission. William's son, William Jr., subsequently sold and donated substantial acreage, including White Rock Mountain, to the state of New York. William Church Osborn and his sons are responsible for creating the Garrison Landing Association which helped to preserve that community in its nineteenth century character. It is now listed as a historic district on the National Register of Historic Places.

Earl D. Osborn, another son of William Church Osborn, gave 1003 acres of land to the state as an expansion of Fahnstock Park in the Highlands. The eldest son of William Church Osborn, General Frederick Osborn, served for over forty years as a commissioner of the Palisades Interstate Park Commission. He was instrumental in persuading the state to acquire Constitution Island Marsh in 1970 and "Olana," the estate near Hudson, N.Y., of the artist Frederic Church, in 1966. He founded and donated land for the Garrison School Forest, a nature center for the local public school, preserving Revolutionary War fortifications (South Redoubt) and the hills above Garrison. His son, Frederick H. Osborn, Jr., and family have been active conservationists and with other branches of the Osborn family recently donated controlling interest in Ardenia Corporation, a 90 acre scenic holding on the Hudson. The grandchildren of both Frederick and William II continued to be active in conservation both in the Hudson Valley and in other parts of the country. Frederick H. Osborn III and Kate Roberts were founding directors of the recently formed Hudson Highlands Land Trust.

Apart from the work of the Osborns, the Highlands continued to provide a creative inspiration for several individuals who in the 1950s and 1960s used this landscape for a new type of expression. Through these efforts thousands of additional acres were preserved, and nineteenth century ideas espoused by Downing were revived. In those years, the Highlands saw the establishment of an outdoor sculpture museum, the relocation and restoration of the Boscobel estate, and the construction of "Dragon Rock" and "Manitoga" as models of the possibilities for design with nature in the context of modern architecture and landscaping. Each of these was bold and different. Each capitalized on the Highlands' sense of place, using the mountain scenery to make a statement on the relationship of the human community to the land. While the roots of their philosophy were a century old, the ideas they explored were thoroughly modern.

"Manitoga" was the project of Russel Wright, one of the best known industrial designers of his time—the first person to use cork, aluminum, and wood in functional household utensils. In 1958 he purchased the abandoned King's Quarry, an 80-acre parcel which was was "uninviting, dry, useless land." Its stone had been used among other purposes for the foundation of the New York Public Library. For the next 25 years, the famous designer turned his energies to renewing its wildness, using the land for an experimental landscape design to create a place where people could come and experience intimate contact with nature in a carefully sculpted setting made to look wild, but in fact carefully planned and executed to create the experience Wright conceived.

Following Downing's tradition, Wright sought to mold the landscape so that the viewer felt a part of nature and experienced its sublime effects, creating from his nondescript, second-growth woodlot and granite pit a wild garden of ferns, vines, wildflowers, and lush forest in which three and one-half miles of paths lead the walker along routes designed to show the land's most interesting features. A sunset and autumn path leads toward the west and includes a place to sit and watch the setting sun. In autumn it shows off the fall colors of wine red viburnum and golden maples set off by giant granite bolders. A path to the "lost pond" stops 100 yards short of the pond itself and leaves the visitor to find the way. Around the curves of paths are offshoots to secret rooms or "alcoves" in the forest, placed so the walker discovers the surprises such as an opening in the forest canopy where dogwood grows over a thick floor of fern.

Selectively cutting trees and brush and allowing others to spread, he carved out vistas of the Hudson framed by large trees, brought out the contrasts of gray birch and dark hemlock, and created small plant communities of native vegetation in a purity of arrangement rarely experienced in nature on a site as small.

For his home, he built "Dragon Rock," a low structure of native white oak and massive boulders, a house of contemporary design. It is built with natural materials to both blend with and contrast with the landscape. It is built on the cliff edge of the old Kings' Quarry. A diverted stream has turned this granite pit into a dark shadowed pool surrounded by ferns, moss-covered stones and hemlock. To Wright and his daughter, the jagged rocks which surround its waters resembled a dragon drinking from a pool so he named the spot and his house "Dragon Rock." The land he named "Manitoga" which means Great Spirit. Like Downing, Wright emphasized the sensuous harmony between natural and human experience. At his death, Wright bequeathed the property to The Nature Conservancy which has set up an independent corporation for its long-term protection.

If Dragon Rock was an experiment with modern architecture, the restoration of "Boscobel" was aimed at re-creating a nineteenth-century landscape and fit into a twentieth-century approach to historic preservation. "Boscobel," a

Portions of "Boscobel," a restored federal style mansion, being put into storage before being reconstructed in the Highlands. Courtesy of Boscobel Restoration.

Federal mansion erected in 1806 by Mr. and Mrs. Staats Morris Dyckman on a thousand acre bluff overlooking the Hudson in Montrose, New York, was sold to a house wrecker for $35 in 1955 to make room for a Veterans Administration hospital. The elegant mansion was considered by many to be an architectural museum piece, inspired by Robert Adam and others, and a group was hastily organized to secure its preservation. To do this, a new location would have to be found, and meanwhile the house was dismantled and stored in barns in Garrison. Eventually, a 36-acre site in Cold Spring, 15 miles to the north of the original location, was chosen for the reconstruction of "Boscobel." The new site in the Highlands, like the original Dyckman estate, commands a view of the Hudson to the south and to the west. The choice of a Highlands location fit into the image of the region as a place where nature and history intertwined. It also preserved a major piece of Highlands scenery and contributed to the momentum of protecting the region from 1950s urbanization and sprawl.

Lila Acheson Wallace, a well-known philanthropist interested in preserving historic sites, gave a half million dollars to reconstruct and furnish the house and landscape the grounds and later contributed $14 million to its preservation and maintenance, including a sizeable endowment. Work began in January 1959 and was completed in 1961. Governor Nelson Rockefeller gave the principal address at a dedication ceremony on May 21. Furnished with the great New York cabinetmakers of the day, the house contains many of the original collections of china, silver and much of the original library. The mansion is situated 200 feet above the Hudson River, opposite West Point, and offers a sweeping view of the Hudson to the south as well as views of Storm King, Breakneck, and Bear

Sculpture by Mark Di Suvero, *Mon Pere, Mon Pere*, 1973–75, in the sculpture garden of the Storm King Art Center in the western Highlands at Mountainville. Photograph by Jerry L. Thompson, courtesy of the Storm King Art Center.

Mountain. Boscobel preserves more than a historic house, it preserves an image of the Highlands.

An entirely different vision was expressed at the Storm King Art Center, where industrialists Ralph E. Ogden and Peter Stern created a 200-acre sculpture garden. Located west of the Highlands' gorge in Mountainville, the art center is at the base of Schunnemunk Mountain, a 2,300 acre preserve which was given to the art center and forms the backdrop for the display of over 100 monumental outdoor sculptures by contemporary artists such as Calder, Noguchi, and di Suvero. Founded in 1960, the art center offers a modern experience of nature and humankind, in which the forms of the landscape are sometimes echoed and sometimes starkly contrasted by the textures, forms, and materials of the sculptures represented. In an ancient setting, with distinctly modern art works, it builds on the reputation of the Highlands as a place where the relationship of the land to our spiritual selves can be explored.

The decade from 1930 to 1940 launched major new conservation efforts in the Highlands. The period of World War II was a time when public agencies and the Hudson River Conservation Society had turned their attention elsewhere, and more individual statements were made through projects such as Manitoga, Boscobel, and the Storm King Art Center. In 1963, new activities at Mount Taurus and across the river at Storm King would bring the Highlands back into public focus.

12. Storm King

IN THE EARLY 1960s, the Nature Conservancy, a national non-profit group which buys outstanding natural areas to assure their long-term protection, identified the northern gate to the Hudson Highlands as an important area for its programs. On September 23, 1962, the Conservancy set up a committee chaired by Leo Rothschild "to look into the acquisition of the Beacon Mountain—Breakneck Ridge—Mount Taurus area," and as Walter Boardman, Executive Director of the Conservancy, later recalled, only the east bank of the Highlands was targeted because Storm King was considered "safe." No one would dream of touching such a well-known landmark.[1] The east bank, however, was still vulnerable to quarrying.

Meanwhile, at Consolidated Edison, Chairman Harland Forbes prepared for a press conference. The utility was about to unveil one of its most ambitious projects: the largest privately owned pumped storage hydro-electric plant in the world. Four days after the Conservancy's first committee meeting, a front-page headline in the *New York Times* announced "Huge Power Plant Planned on Hudson." The *New York Times* article reported that the location would be "near Cornwall, New York."[2] Forbes was quoted as saying that "no difficulties are anticipated." The location was Storm King Mountain, and Forbes could not have been farther from the mark.

Within months, a seventeen-year legal battle began that today is credited with launching the modern environmental activism. The case would set far-reaching precedents for conservation: It would address the protection of natural resources and scenic beauty as a public purpose equal in importance to energy development; it would establish the right of citizen groups to sue a government agency to protect such resources; and it would lead to the passage of the National Environmental Policy Act.

The Storm King issue pitted industry against conservation in a dramatic clash over an area which was well suited for both. The clash was the culmination of a process that began in 1908, when the Harrimans blocked the construction of a prison in the Highlands, creating a public park instead. The Storm King case would involve the children and grandchildren of a number of the captains of industry who had settled the Hudson Valley in the 1880s, and it would eventually split some of them in a divisive ideological war. The fight to protect the mountain would also draw together thousands of ordinary people from nearly all American states and several foreign countries, pitting them against a utility convinced that its plan for a power plant was good not only for the public but

CON EDISON'S PROPOSED HYDROELECTRIC PROJECT, CORNWALL, NEW YORK

This rendering shows how Con Edison's proposed pumped storage hydroelectric project will look upon completion in 1967.

At times when power needs are high, water will flow from the reservoir down through a 40-foot diameter, 2-mile long tunnel to the plant where it will drive turbines attached to generators having a maximum capacity of 2,000,000 kilowatts. The electricity will be transmitted at 345,000 volts by submarine cables under the Hudson River and then by overhead lines into the system. During the low load periods the procedure will be reversed and water will be pumped from the river up through the tunnel into the reservoir.

The turbine-generators are of the reversible type and during pumping periods the generators will operate as motors, the turbines as pumps. Power for pumping will be supplied from the Company's steam generating units in New York City and Westchester County.

The nature of Con Edison's growing electric load makes the economics of a pumped storage hydroelectric plant highly favorable. The greatest demands for electricity occur during the daytime hours on weekdays. Demands during nights and weekends drop to less than one-half, with the result that much of the new efficient steam generating capacity is not being anywhere near fully utilized. When the pumped storage project is completed, however, low cost electricity from these units will be sent northward over the lines to pump water at night and on weekends so as to make available substantial added capacity to meet weekday daytime loads. Thus the reservoir functions as a kind of huge storage battery.

The chart illustrates how the Cornwall station might be operated during a typical 24-hour period on a summer weekday. From 8 A.M. to 7 P.M. the station would be in operation supplying electric power to the system. Around midnight, as electric power from the efficient steam-generating units on the system became available for pumping, the process of refilling the reservoir would start, to continue until 7 A.M. the next morning.

17

Artist's rendering from Con Ed 1962 Annual Report. Courtesy of Consolidated Edison, New York.

also for the scenery. The ranks of the opposition would include stockholders of Con Ed who would donate their dividend checks to fund the lawsuit against the company. The court battles would result in major losses for both sides, leaving the fate of the mountain hanging in the balance for nearly two decades.

The conflict began in 1961, when Con Ed was approached by its neighboring utility, Central Hudson, with a plan for a "pumped storage" power plant. Pumped storage, a new technology conceived as a way to provide power during hours of peak demand, operated with excess electricity available during off-peak hours to pump river water through pipes up a hill or mountain to a storage reservoir. The water would be released when electrical power was most needed, during peak hours, powering a reversible turbine as it coursed down the mountain. A drawback was that it would take three kilowats of electricity to pump the water up to the storage reservoir, from which, when the water was released, two kilowatts of power would be generated. The advantage was that the power would be available immediately, on demand. In addition, new power plants, which would have been idle much of the time, would not have to be built.

Changing patterns of electrical consumption were forcing all Northeast utilities to plan for peak-hour needs. According to one source, in the early days of power generation, peak-load often occurred on Christmas Eve, when tree lights and toy trains placed the greatest strain on the power system.[3] By the 1960s peak loads were occurring on hot summer days, when air conditioners were running at full blast, and during the long nights of winter, when lighting needs were greatest.

Storm King Mountain and Breakneck Ridge were logical sites for pumped storage power plants, Central Hudson officials told Con Ed, adding that they planned to build on Breakneck. Central Hudson suggested that the Storm King site, which would require more capital and produce more power, might be of interest to the larger, New York-based utility, and they offered to buy some of the power generated by the Storm King plant if Con Edison decided to proceed.

Con Ed investigated the Storm King site, and by the fall of 1962, some of the necessary land had been purchased and plans had been drawn to develop it. Con Ed submitted a license application to the Federal Power Commission for a two-million kilowatt plant costing $115 million, and announced its project to the press. Meanwhile, Central Hudson was also drawing up its plans and negotiating an option on 660 acres on Breakneck Mountain.

From the utilities' standpoint, Breakneck and Storm King were ideally situated for pumped storage power plants. In one of its promotional brochures, Con Ed stated:

"At Cornwall, 50 miles from habitation, nature has provided the physical requirements of a pumped storage hydro-electric station: a large source of water and a reservoir basin high above but near enough to store water for generating power. Selected after years of intensive study and consideration of possible alternatives, the Cornwall site is one of the best in the world for such a hydrostation."[4] Such a plant at Storm King would serve as a "gigantic storage battery for the Con Edison system," company respresentatives said.

What Con Ed did not know was that it had chosen a site which many Americans regarded as sacred ground. The battle to save it soon assumed the intensity and proportions of a holy crusade. The Northern Gate, the Hudson River Gorge, the Wey-gat—by any of these names the passage through Breakneck and Storm King had come to symbolize the essence of the Highlands and the Hudson Valley region. This place of contrast, of confined river with broad valley, of shadow with light, and of river, mountain and sky had captured the imagination of the artists of the Hudson River School. The scene of the Northern Gate, where the river meets the mountains, was painted and sketched more than any other scene in the Highlands. This was the view that nineteenth-century tourists saw from the plains at West Point; it greeted them from sloops and steamers as they journeyed south across the broad expanse of Newburgh Bay; and it is the image that people today are most likely to remember. With Bannerman's Castle at its base, the Northern Gate is a threshhold to a world of singular beauty.

So it is not surprising that the threat of a power plant on each of the two sentinels of the Northern Gate provided the focus for a great national outpouring of emotion, effort and funds over a period of almost two decades. Not merely a power plant but a principle was at stake. Storm King embodied the history of a nation and its connection with the land. This was a fight to protect a place of symbolic value. Ultimately, although compromise was offered, none was possible.

Con Ed's news release, which referred to a site in the "Cornwall area," did not immediately trigger a public outcry, because the exact location was not generally known. However, a briefing soon after, by Con Edison's Vice-President for Engineering, Motton Waring, roused the grave concern of the Palisades Interstate Park Commission. The power plant, they learned, was to be built between Crow's Nest and Storm King on PIPC park land. The reservoir would be built nearby on lands owned by the town of Cornwall and on Black Rock Forest, owned by Harvard University. The engineering plans called for transmission lines to be strung across the Storm King gorge from tall towers.

The PIPC Commissioners were some of the leading conservationists in the Hudson Valley. Among them were Laurance S. Rockefeller, the Governor's brother, and Frederick Osborn, a descendent of William Henry Osborn who had built "Castle Rock" in Garrison. The Commissioners expressed concerns about the wisdom of constructing a power plant on park land, and they cautioned against destroying the scenic vista of the gorge. Osborn also discussed the project with his brother William, who was president of the Hudson River Conservation Society. The PIPC was soon joined in its objection by General Westmoreland, Commandant at West Point, who was concerned that the transmission lines might interfere with helicopter traffic en route to the academy and by the Hudson River Conservation Society.[5]

Eager to eliminate any political obstacles, Con Ed's Vice President Waring responded by agreeing to move the power plant off the park land onto the north face of Storm King and to put the transmission lines underground where they

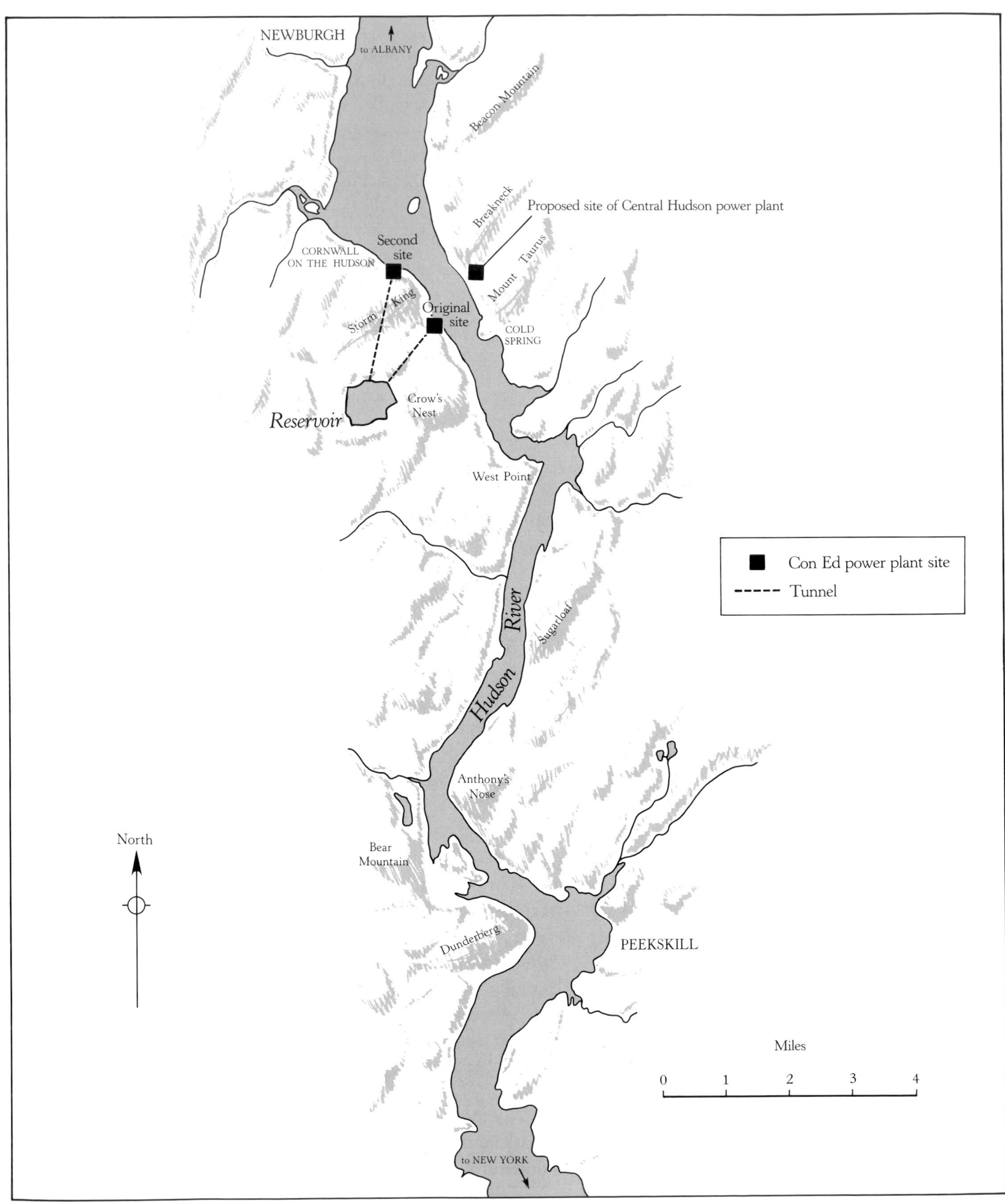
NEWBURGH
to ALBANY
Beacon Mountain
Breakneck
Proposed site of Central Hudson power plant
Second site
CORNWALL ON THE HUDSON
Mount Taurus
Storm King
Original site
COLD SPRING
Reservoir
Crow's Nest
West Point
Con Ed power plant site
Tunnel
Hudson River
Sugarloaf
Anthony's Nose
North
Bear Mountain
Dunderberg
PEEKSKILL
Miles
0
1
2
3
4
to NEW YORK

crossed the Hudson. The lines would, however, continue to be above ground on the east side of the Hudson across Putnam County to New York City. The added cost to Con Ed would be about $6 million, but it would grease the skids for construction of the plant. After these concessions, PIPC and the Conservation Society dropped their objections to the plant, as did Westmoreland and West Point.

The PIPC and the Conservation Society were soon to discover that they were out of step with the thinking of many other conservationists in the Hudson Valley. In April 1963, the company published its annual report, showing the revised plan with the new Storm King location. If anyone doubted the impact of a power plant on the mountain scenery, the report graphically dispelled any such thoughts. It included an artist's rendering of the proposed plant, showing Storm King Mountain with a cavernous, square hole carved out of its northern face. This opening was filled with transformers and switch yards, from its base at the river shore to the height of the Storm King highway. The moment the report was published the battle cry was sounded.

Leo Rothschild was the first to express alarm over Con Ed's proposal. In addition to heading up the Nature Conservancy committeee, Rothschild was an avid hiker and served as Conservation Chairman of the New York-New Jersey Trail Conference, a group which maintained a system of hiking trails on Storm King near the proposed reservoir location. Soon after Con Ed published its annual report, he wrote a telegram to Governor Nelson Rockefeller, citing the Governor's interest in scenic and historic landmarks, and calling on him to do everything in his power to preserve the Highlands.[6] In this message, he deplored "the desecration of the northern gate of the Hudson Highlands." Rothschild copied his telegram to the *New York Times*. Central Hudson's announcement shortly after that it planned a similar power plant at Breakneck only added to his concern.

Rothschild was soon joined in his efforts by Carl Carmer, the New York State Historian and author of a book about the Hudson, as well as Walter Boardman from the Nature Conservancy and others. Following the publication of Con Ed's annual report, they spent several months trying to persuade officials at all levels to relocate the project. They spoke to Waring at Con Ed; they wrote to Stewart Udall, Secretary of the Interior; they sent a delegation to meet with Governor Rockefeller; and they wrote to his brother Laurance.

Their pleas fell on deaf ears. Waring remained committed to construction at Storm King. Nelson Rockefeller suggested that if they were so concerned, they should buy the mountain. Laurance Rockefeller stood by his decision as a PIPC Commissioner to endorse the project. Although Laurance Rockefeller was one of the country's most outspoken leaders in preserving natural beauty, he was also a proponent of multiple use of natural resources. He felt that with mitigation of the scenic impacts, the plant was an appropriate use of the area.[7] His backing helped guarantee the support of his brother.

Only the *New York Times* offered support and published an editorial titled "Defacing the Hudson." It read:

Sites of proposed power plants.

> If any utility proposed to construct a plant in the middle of Central Park, the absurdity of such a defacement of precious natural (or nearly natural) surroundings would be immediately apparent. It is almost as bad to plunk down a couple of power installations right in the heart of one of the most stunning natural regions in the Eastern United States: Storm King Mountain (north of Bear Mountain and West Point) and Breakneck Ridge on the opposite (eastern) side of the Hudson.
>
> All of us who have driven down from New England and northern New York have looked with awe at these breath-taking mountains. All of us who have hiked and played in Palisades Interstate Park know what a beautiful backyard exists 50 miles north of New York. Is it too close to home to appreciate? 'This is very good land to fall with and a pleasant land to see,' said one of Henry Hudson's officers, going up the river under these high blue hills. That great traveler Baedeker found the Hudson's scenery 'grander and more inspiring' than the Rhine's.
>
> The proposed power plants of Consolidated Edison at Storm King and of Central Hudson Gas and Electric at Breakneck Ridge would desecrate great areas that are part of the natural and historic heritage of our country, are still largely unspoiled and should remain that way.[8]

The *Times* editorial was encouraging, but it held no sway with the utilities. Having failed on the diplomatic front, opponents of the power plant resolved to pursue their cause in court. On November 8, 1963 a small group of people met at the home of Carl Carmer in Irvington and formed the Scenic Hudson Preservation Conference—a coalition of conservation groups, including the Nature Conservancy and the New York-New Jersey Trail Conference—to oppose the power plant.

It was an interesting alignment of power and philosophies with many built-in ironies. Supporting the power plant were some of New York's most committed conservationists, Laurance Rockefeller and the two Osborns, all deeply interested in preserving the Hudson's scenery, and through their actions, carrying on a tradition established by their grandparents. Also supporting the power plant were two highly respected institutions, the Palisades Interstate Park Commission and, briefly, the Hudson River Conservation Society.

On the opposing side were others with equally impressive credentials: author and historian Carl Carmer; Walter Boardman, then director of the Nature Conservancy; and Leo Rothschild of the New York-New Jersey Trail Conference. Initially, only two Hudson Highlands residents spoke out against the project: Ben Frazier, a Hudson Highlands resident who was particularly knowledgeable about the area, and Alexander Saunders, a man who owned a manufacturing business in the Nelsonville area near Cold Spring. A year later, about twenty of the "weekenders" who owned country homes on Storm King joined the fight, among them Chauncey Stillman, grandson of James Stillman, and the Duggan family.

Ironically, James Stillman, J. P. Morgan and William Rockefeller had been creators of Con Edison's precursor, the Consolidated Gas Company, a monopolistic holding company for virtually all of New York City's gas and electric utilities, including Edison Electric. Eighty years later, the company they created was proposing a power plant on the face of Storm King, on land Morgan had attempted to set aside as "forever wild" by making it part of New York State's Forest Preserve. Nearby was the mountain retreat in Cornwall established by Stillman to form the basis for his art colony/resort. His descendants rallied together to prevent Con Edison, once a source of their family wealth, from building on the mountain scenery they cherished. So committed were the Stillmans to the preservation of the area that they had donated land on Storm King to the Palisades Interstate Park Commission and had given the 3600-acre Black Rock Forest to Harvard University with an endowment for its operation as an experimental forest. The Duggans' relatives, similarly, had donated to the Village of Cornwall the lands for its reservoir and watershed, the same lands which would now be purchased from the Village by Con Ed for a project the Duggans opposed.

With Carl Carmer as Chairman and Walter Boardman as President, the newly formed group decided that the Federal Power Commission (FPC) licensing hearings presented the only opportunity for a legal challenge. Con Ed had filed a license application with the FPC in January 1963 and amended it in September. License hearings were scheduled for February 1964, and Scenic Hudson could try to persuade the FPC to deny the license. The same approach could be tried for the plant on Breakneck, once Central Hudson filed for a license. It was a meager hope at best. The FPC conducted primarily technical and economic reviews, and generally ignored social values.

Scenic Hudson board members began by searching for an attorney with experience in FPC proceedings who might be sympathetic to their cause. Three weeks before the the February hearings began, they persuaded Dale Doty to take the case. Doty, a Californian, was a former FPC Commissioner who had built a law practice around representing companies before the agency. His record as an FPC Commissioner showed a concern for social policy and the environment. For example, at his urging, the FPC, in a highly unusual move, had refused to license a proposed hydroelectric dam which would have destroyed the recreational use of the Namekagon River in Wisconsin. The Storm King case intrigued Doty because of the scenic and conservation policy issues it raised and the opportunity it presented to broaden the view of public purposes to be protected by FPC in licensing a power plant.[9]

The FPC hearings began on February 25, 1964, and Doty, with only three weeks to prepare, filed for Scenic Hudson to intervene in the proceedings. His first step was to stall for time to prepare his case, and this request, coming from a highly respected former Commissioner, was granted. The FPC gave Scenic Hudson until May to assemble witnesses and review the technical evidence presented by Con Ed. Doty then spent the next few months preparing the legal arguments against a power plant on the basis of scenic and historic considerations. He sought to demonstrate that construction on Storm King was not the only way to meet Con Ed's peak power needs. Using technical and economic arguments, Doty hoped to prove that Con Ed could use other equally viable methods which would spare the mountain.

The May hearings opened with presentations from both sides. Doty called several witnesses to support his case, among them a former FPC engineer who argued that gas turbines could serve as an alternative to pumped storage. A hydrologist testified on possible seepage problems. However, Doty's primary witnesses addressed the scenic and historic importance of Storm King and the Highlands. Ben Frazier eloquently described the region's role in American history, beginning with the great chain across the river in the American Revolution, continuing with the fortification of West Point, the Civil War ironworks at Cold Spring, and concluding with a description of Storm King's grandeur. Doty, in his summary, reviewed for the hearing examiners how the beauty of these mountains had influenced the Hudson River School of Painting and the romantic revival in landscape design, architecture and literature. Leo Rothschild told of Storm King's recreational importance for hikers, and Carl Carmer, the author/historian, argued that the Highlands were significant as a national natural heritage and should be kept free from the invasion of industry. His testimony framed the issue as it would be debated for nearly two decades:

> Is it fitting, then to ask the question whether this beauty shall be sacrificed to these enterprises which would change the shoreline, lower high peaks, destroy groves of trees. . . . The Hudson answers a spiritual need more necessary to the nation's health than all the commercial products it can provide, than all the money it can earn. . . . We believe that ugliness begets ugliness and that nature's beauty, once destroyed, may never be restored by artifice of man. . . . We would offer the peace and healing our river gives, as it has always given, to those who seek its waters for respite from the tension of their lives.[10]

The opposing view was offered by Randall Le Boeuf, Con Ed's lead attorney, who presented a primarily technical case to the FPC. He outlined the need for the project, the cost-effectiveness of pumped storage, and the operating efficiencies to be achieved.

Le Boeuf argued that the plant would be a wise investment for the utility. First, it would provide peak power and emergency service at the lowest cost; it could provide two million kilowatts at about eighty percent of the cost of a conventional fossil fuel or nuclear plant.

Second, it would improve the efficiency of Con Ed's operation, since a conventional plant is economical only when operated on a round-the-clock basis. By providing power to meet peak loads, it would operate more economically.

Third, it would improve the reliability of Con Ed's entire generating and delivery system. At the time, Con Ed was faced with monthly breakdowns of its steam turbine generators, resulting in outages of 1500 megawatts. Pumped storage plants, Con Ed said, were simple, rugged, and nearly trouble-proof. They could bring power back on line much faster than conventional power plants. In addition, the "Cornwall Project" would allow Con Ed to retire some of its older power plants and provide increased capacity for electric home heating, which the company was beginning to promote.

The Storm King site, Le Boeuf explained, was "uniquely advantageous." It was close to Con Ed's load centers, on a tidal river whose flow would not be affected, and in an area geologically suited for a reservoir and power plant. With the reservoir at 1160 feet above the powerhouse elevation, it offered a "developable head" of enormous value.

Responding to Scenic Hudson's criticisms, Le Boeuf offered the opinion that the aesthetic qualities of the area would be improved when Con Ed razed the decayed piers and buildings along Cornwall's blighted waterfront. In addition, Con Edison would create a mile-long waterfront park. He also suggested that Con Ed's structure might eventually assume the same historic significance to future generations as had the Cold Spring Foundry. He reminded the Commission that Con Edison had made design changes so that the powerhouse would be less visible and noted that the utility, at considerable cost, had agreed to cross the gorge with submarine electric cables, instead of stringing them above the river as originally planned. No transmission lines would be visible from the river. "As a result, the power plant will be handsome in design, adorning the waterfront below Storm King Mountain and replacing the clutter of structures that now mar the landscape of the west bank of the river at the project site." Le Boeuf dismissed Scenic Hudson's testimony as "self-centered complaints" and portrayed it as the backyard protectionism of a few local dreamers.[11]

The Village of Cornwall also testified in support of the project at the FPC hearings. The Village, stood to gain more than half a million dollars in tax revenue from the plant, over half of the Village budget; and Village residents, also standing to gain from the power plant, overwhelmingly endorsed it in a vote of 499 to 25. Mayor Donahue spoke before the FPC on the benefits to the town of added tax revenue and the recreational benefits associated with the park Con Ed would build.

On July 31, 1964, the hearing examiner, Edward B. Marsh, submitted his recommendation to the FPC Commissioners. His conclusion was that the plant should be approved. The final decision rested with the full Commission, which scheduled its review of the proceedings and final hearings for November 1964.

In the meantime, however, support for the preservation of Storm King was spreading across the country. The issue brought into sharp focus a growing national concern over the loss of a natural heritage. As the FPC prepared for the November hearings, the story of Storm King appeared with increasing frequency in the news media. In July 1964 *Life* magazine published an editorial: "Must God's Junkyard Grow?" which expressed those fears.

"When does a local conservation issue become national?" the *Life* editorial asked.

> Up the river near West Point, Consolidated Edison proposes building the nation's third largest hydroelectric station in a hole gouged out of the flank of Storm King Mountain. This is the gateway to the Hudson Highlands, one of the grandest passages of river scenery in the world . . .

> National or local? Apparently local . . . like the battle to save the Indiana Dunes on Lake Michigan from commercial exploitation. Or like the battle to keep California's Kings and Tehipite canyons from being dammed and drowned. But like these and other campaigns . . . they also deserve to be called national issues. In a real sense, the whole country suffers every time Americans make a bad choice. . . . The destruction of such resources is irrevocable; no one can pass that way again.[12]

Such press coverage convinced Scenic Hudson that there was strong support for its position. Its leadership decided to make its arguments heard in the court of public opinion. With the hearing officer in favor of granting the license, this seemed to be the only way to sway the full Commission. Scenic Hudson turned to a public relations firm, Selvage and Lee, to put a spotlight on the case. James Cope, the chairman of the firm, and his assistants, Lou Frankel, Ray Baker and William ("Mike") Kitzmiller, went to work with a zest. They masterminded a highly effective strategy to take the Storm King issue out of the narrow confines of the FPC hearing, where it was undergoing mainly technical review, and expose it to public debate on scenic and historic terms.[13]

Their efforts gave the project a high profile in the media and kept it there. For starters, in September 1964 Cope, Kitzmiller, and a group of Scenic Hudson supporters organized a "waterborne picket line" to protest Con Ed's plans. A regatta of fifty boats, led by the *Westerly*, the "flagship of the New York Yacht Club," assembled at Garrison and set off for Cornwall. When they reached Storm King, two Garrison residents, Lars and Dean Anderson, jumped ashore attired in Colonial uniforms, lit flares, and on the site of the proposed hydroelectric plant, placed a sign stating "Dig You Must Not." (Con Ed's motto at the time, emblazoned on all of its orange trucks, was "Dig We Must.") As the regatta prepared to depart, a telegram arrived from Senator Jacob Javits commending the effort—"Greetings to Commodore Stillman and all his guests aboard on the 'Westerly' today," it stated, ". . . The beauty of the Hudson River and Valley is one of our state's most valuable resources, and all citizens should be willing to help preserve it." The event was photographed and reported in all of New York's newspapers (it then had seven), as well as in *Newsweek*. In Cornwall, where the project would be located, a front-page picture in the local paper showed the Anderson brothers, muskets smoking, with a caption explaining that this marked the spot where Revolutionary armies scouted the invading British.[14]

Around this time, Scenic Hudson, at Selvage and Lee's urging, hired its own Executive Director, Rod Vandivert. Scenic Hudson retained Selvage and Lee as public relations counsel and Vandivert worked in the firm's offices for a number of years. He became the organization's spokesman and, with an active group of volunteers, ran its daily operations. He was later joined by assistant Nancy Mathews, who was also active in the Sierra Club.

Through Vandivert and Kitzmiller's efforts, Scenic Hudson also got the attention of Republican State Senator, R. Watson Pomeroy, Chairman of the New York State Legislative Committee on Natural Resources. Pomeroy held independent hearings on the project which allowed witnesses to be heard who

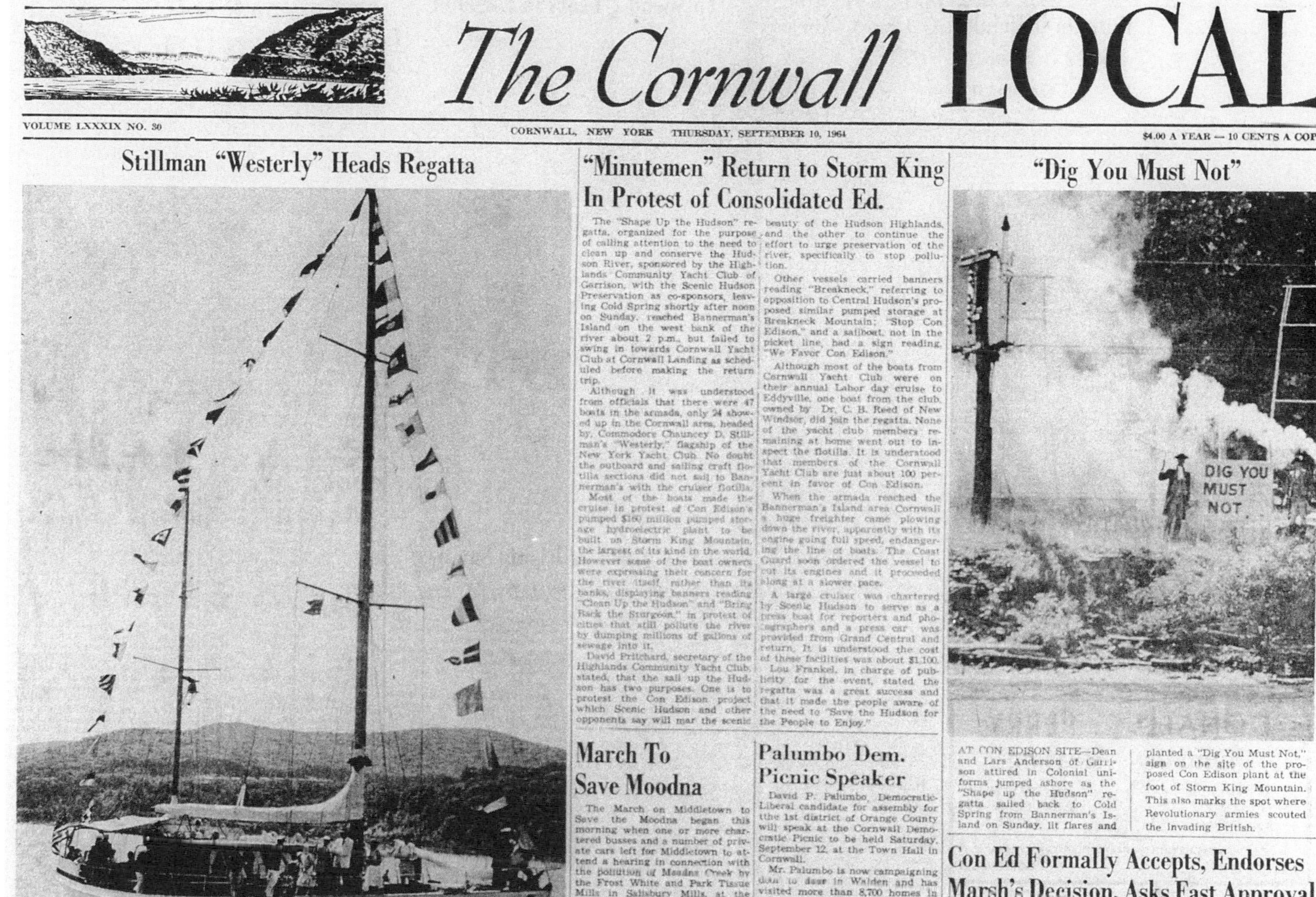

The Cornwall LOCAL

VOLUME LXXXIX NO. 30 CORNWALL, NEW YORK THURSDAY, SEPTEMBER 10, 1964 $4.00 A YEAR — 10 CENTS A COPY

Stillman "Westerly" Heads Regatta

"Minutemen" Return to Storm King In Protest of Consolidated Ed.

The "Shape Up the Hudson" regatta, organized for the purpose of calling attention to the need to clean up and conserve the Hudson River, sponsored by the Highlands Community Yacht Club of Garrison, with the Scenic Hudson Preservation as co-sponsors, leaving Cold Spring shortly after noon on Sunday, reached Bannerman's Island on the west bank of the river about 2 p.m., but failed to swing in towards Cornwall Yacht Club at Cornwall Landing as scheduled before making the return trip.

Although it was understood from officials that there were 47 boats in the armada, only 24 showed up in the Cornwall area, headed by Commodore Chauncey D. Stillman's "Westerly," flagship of the New York Yacht Club. No doubt the outboard and sailing craft flotilla sections did not sail to Bannerman's with the cruiser flotilla.

Most of the boats made the cruise in protest of Con Edison's pumped $160 million pumped storage hydroelectric plant to be built on Storm King Mountain, the largest of its kind in the world. However some of the boat owners were expressing their concern for the river itself, rather than its banks, displaying banners reading "Clean Up the Hudson" and "Bring Back the Sturgeon," in protest of cities that still pollute the river by dumping millions of gallons of sewage into it.

David Pritchard, secretary of the Highlands Community Yacht Club, stated, that the sail up the Hudson has two purposes. One is to protest the Con Edison project which Scenic Hudson and other opponents say will mar the scenic beauty of the Hudson Highlands, and the other to continue the effort to urge preservation of the river, specifically to stop pollution.

Other vessels carried banners reading "Breakneck," referring to opposition to Central Hudson's proposed similar pumped storage at Breakneck Mountain; "Stop Con Edison," and a sailboat, not in the picket line, had a sign reading, "We Favor Con Edison."

Although most of the boats from Cornwall Yacht Club were on their annual Labor day cruise to Eddyville, one boat from the club, owned by Dr. C. B. Reed of New Windsor, did join the regatta. None of the yacht club members remaining at home went out to inspect the flotilla. It is understood that members of the Cornwall Yacht Club are just about 100 percent in favor of Con Edison.

When the armada reached the Bannerman's Island area Cornwall a huge freighter came plowing down the river, apparently with its engine going full speed, endangering the line of boats. The Coast Guard soon ordered the vessel to cut its engines and it proceeded along at a slower pace.

A large cruiser was chartered by Scenic Hudson to serve as a press boat for reporters and photographers and a press car was provided from Grand Central and return. It is understood the cost of these facilities was about $1,100.

Lou Frankel, in charge of publicity for the event, stated the regatta was a great success and that it made the people aware of the need to "Save the Hudson for the People to Enjoy."

March To Save Moodna

The March on Middletown to Save the Moodna began this morning when one or more chartered busses and a number of private cars left for Middletown to attend a hearing in connection with the pollution of Moodna Creek by the Frost White and Park Tissue Mills in Salisbury Mills, at the Middletown District Office of the State Health Dept.

Palumbo Dem. Picnic Speaker

David P. Palumbo, Democratic-Liberal candidate for assembly for the 1st district of Orange County will speak at the Cornwall Democratic Picnic to be held Saturday, September 12, at the Town Hall in Cornwall.

Mr. Palumbo is now campaigning door to door in Walden and has visited more than 8,700 homes in the 1st assembly district.

Home Unit To

"Dig You Must Not"

AT CON EDISON SITE—Dean and Lars Anderson of Garrison attired in Colonial uniforms jumped ashore as the "Shape up the Hudson" regatta sailed back to Cold Spring from Bannerman's Island on Sunday, lit flares and planted a "Dig You Must Not," sign on the site of the proposed Con Edison plant at the foot of Storm King Mountain. This also marks the spot where Revolutionary armies scouted the invading British.

Con Ed Formally Accepts, Endorses Marsh's Decision, Asks Fast Approval

Con Edison has filed a brief with the Federal Power Commis-

they said had been unduly overblown because of the zeal of a

"Dig You Must Not"—Citizen protest organized by Scenic Hudson. Courtesy of *The Cornwall Local.*

had not testified before the FPC. Among these witnesses were Alexander Lurkis and Robert H. Boyle. Lurkis, a paid consultant, testified about gas turbines as an alternative source of peak power, giving further credence to the testimony presented by Scenic Hudson witnesses in May. Boyle, an editor and writer for *Sports Illustrated* and an avid Hudson River sport fisherman and naturalist, testified that Con Ed's nuclear plant at Indian Point, 15 miles downriver, was already destroying fish in the river and that the Storm King plant would add to the carnage. He pointed out that the Hudson Highlands was a major spawning ground for Atlantic coast striped bass and suggested that the stripers' eggs, larvae, and fingerlings would be sucked into the power plant with the intake water and destroyed. Another witness, John R. Clark, of the U.S. Marine Gamefish laboratory at Sandy Hook, New Jersey, spoke on the importance of Hudson-spawned striped bass to coastal waters.

The legislative hearings put additional pressure on the FPC, which delayed a final decision for several more months. On February 16, 1965, Senator

Pomeroy's committee released its preliminary report. It stated: "On the basis of information which has come to this committee's attention, we find that substantial questions of great importance to the people of New York remain unanswered regarding the Con Ed project. . . . this committee must go on record as being opposed to its approval."[15]

At this time, the Hearing Officer for the FPC issued his recommendation that the plant be licensed, but before the FPC acted on his report, the people at Selvage and Lee organized a campaign in which a number of groups petitioned the FPC to reopen the proceedings to consider new evidence, and also to admit them as parties. The new evidence was primarily the detailed testimony of Alexander Lurkis and the information on striped bass spawning at Storm King which came out at the Pomeroy hearings. The striped bass claim was further substantiated by a previously undiscovered study by Rathjen and Miller, who had actually netted stiped bass eggs in the Hudson and had found more of them at West Point than any other place. This information, plus the Lurkis report was formally offered to the FPC, before it reached its decision to license, in petitions to reopen filed by about ten groups such as the Hilltop Cooperative Apartments (who claimed that the Lurkis alternative would save them money) and the National Partyboat Owner's Alliance, who claimed that destruction of the striped bass would cause them irreparable injury. All of these petitions were denied by the FPC, which asserted that they came too late and didn't add anything, although there had been nothing in the FPC hearing record as detailed as Lurkis' proposal on gas turbines presented in his Pomeroy testimony, and nothing at all about the striped bass spawning at West Point.

On March 9, 1965 the FPC completed its review and issued its decision. The license to Con Edison was granted. In responding to the issues raised by Scenic Hudson at the hearings, the FPC concluded: "Here the impact of the project on the surroundings is minimal while the need for electricity from this economical and dependable source is great. We therefore conclude that on balance the issuance of a license . . . is desirable and justified in the overall public interest."[16]

Scenic Hudson's only option was to appeal, and at this critical moment, significant help was offered by the Taconic Foundation. They agreed to finance Scenic Hudson's appeal if the group would use the services of Lloyd Garrison, a partner in the law firm of Paul, Weiss, Rifkind, Wharton and Garrison, and a board member of the foundation.[17] At this point Doty, who had artfully laid the groundwork for such an appeal in his development of the hearing record, turned the case over. Lloyd Garrison drew on the assistance of several other attorneys in the law firm and assembled a team which included two of the partners and several new staff members who had recently graduated from Harvard and Yale law schools. Among them was Albert Butzel, who was to work on the case for its seventeen-year duration, eventually bringing it with him to his own firm, Berle and Butzel.

When the FPC swept these offers of new evidence aside and licensed the plant notwithstanding, it set the stage for an ingenious legal approach. The

Federal Power Act required the FPC to consider applications for hydroelectric power plants in the context of a comprehensive waterway plan which addressed both power development and "other beneficial public uses, including recreational purposes," which, previous cases had determined, included natural beauty.[18] Because of this, Garrison planned to argue that, given the recognized scenic beauty of Storm King and the Highlands, the FPC should have bent over backward to seek out feasible alternatives. Instead they had turned their backs on the offer of the Lurkis proposals, and had not undertaken themselves to investigate other possible alternatives such as the use of interconnected powers. Garrison and his staff made essentially the same point with respect to striped bass—that what little evidence there was on fish in the record was meaningless and that when the FPC was presented with new evidence indicating the potential of a real threat to the striped bass, they were duty-bound to inquire into it. Instead, Scenic Hudson argued, the FPC dismissed the petitions as filed too late. The case was presented to the Second Circuit Court, which scheduled the appeal for June 1965.

As Scenic Hudson began preparing its appeal, public pressure for protection of Storm King was mounting, and with it concern for the protection of the entire Hudson River. The Bear Mountain hearings held by Senator Pomeroy drew dramatic attention to the fact that despite decades of abuse, the Hudson estuary still cradled millions of young shad and striped bass, future migrants to the Atlantic coast fishery. The hearings brought to light the massive fish kills then occurring at Con Ed's Indian Point nuclear reactor and illuminated the conflicts over the future uses of the Hudson for power, water supply, commercial fishing, historic preservation, and recreation. Most of all, it showed the fragile protection accorded to the Hudson's natural resources.

In the 1960s, as now, the Hudson Valley's population was swelling at a rapid rate, threatening to intensify conflicts and leading to the possibility of irreversible destruction of the Valley's scenic beauty, wildlife, and water supplies. With articles about the Hudson and Storm King appearing frequently in the state and national press, the issue of protecting the Hudson quickly began echoing in the halls of Congress and in Albany. The battle had moved out of the FPC hearing room into the political arena.

Running in 1964 on a platform of river protection, Richard Ottinger had captured the Congressional seat of a traditionally Republican Westchester and Putnam County district, an area that would be traversed by the powerline from the Storm King plant. Once in office, he hired Mike Kitzmiller, Scenic Hudson's public relations consultant, as his legislative aide, and began to propose legislation for the Hudson River, its fishery and the Storm King project. Congress was to hold a number of hearings on Ottinger bills in the next five years, and the new Congressman used them as a forum to attack Governor Rockefeller's record on the Hudson. Rockefeller responded to these atacks by announcing new programs for the river. The two politicians thus kept the Hudson in the limelight for several years.[19]

Following the first of these hearings, the Congressman introduced legislation for a "Hudson River Scenic Riverway." The bill was picked up in the Senate by both of New York's Senators, Republican Jacob Javits and Democrat Robert F. Kennedy, as well as Harrison Williams, Democrat of New Jersey. In the House of Representatives, Congressmen Jonathan Bingham, Ogden Reid and many others joined forces with Ottinger in support of the legislation. The Riverway bill, prompted by the Storm King case, would give the U.S. Department of Interior $20 million to acquire land and conservation easements in a mile-wide corridor on each side of the river for a distance of fifty miles, from Yonkers north to Newburgh, including the Highlands and encompassing territory in the states of New York and New Jersey. The purpose of the Riverway would be to provide for public recreation on the Hudson and its shores and to preserve the historic, scenic and natural resources of the river corridor. The bill stated that the Federal Power Act "shall not operate to affect adversely" the Riverway.[20]

It was a direct jab at New York State for failing to properly protect the Hudson River, and Governor Rockefeller voiced his opposition to it. Undeterred, Congress pressed its case. Senator Jacob Javits, speaking before Congress on March 4, 1965, expressed the view of many of his colleagues:

> I believe in the philosophy of government known as the Lincolnian belief or the Lincolnian theory, which should dominate my party. The essence of that philosophy is that when a State cannot do for itself as well or at all what needs to be done in the interest of the whole Nation, the Nation itself must do it.
>
> In this instance, I do not favor delay. That is why I have joined, currently, in the bill introduced by the junior Senator from New York.[21]

The junior Senator, Robert Kennedy, speaking on the same day, added his own eloquent reason for action:

> Consideration of this bill is essential because of pending encroachments on the Hudson and a heightened national awareness of the need to preserve those areas constituting our national heritage. Preservation of part of the area surrounding the lordly Hudson is of particular concern to our citizens because of the central role it has played in the history of our country, because of its unique beauty and majesty, and because of the important part it plays in the lives of such a large number of our citizens. . . . Many foreign visitors coming to the United States ride up the Hudson to get their first impressions of the American countryside. Of even more importance are the countless number of future Americans who will judge us by our actions in preserving areas such as the Hudson River Valley. They rely on us to insure that "America the Beautiful" is a fact as well as a phrase.[22]

Ottinger, the lead sponsor of the Riverway bill, also spoke on the record and took shots indirectly at the Governor:

> Mr. Speaker, I regret to say that one of the Nation's greatest rivers has not yet benefited from this recent awakening of official concern. In spite of unparalleled public demand, in spite of the efforts of the State and Federal legislators and a wealth of proposals for its development and restoration, one

> river has been exempt from constructive official action. I am speaking of the great and historic Hudson River—without question one of the Nation's most beautiful and valuable river resources.
>
> "It is generally recognized that no river in America is in greater need of help than our beloved Hudson: it stands of the brink of destruction. Yet it also stands on the threshold of a great new future. Decisions will be made in the next few months which will determine whether it will become a model of river development or an industrial canal.[23]

Not to be outdone, three months after Ottinger submitted his bill, Rockefeller issued an Executive Order appointing his own Hudson River Valley Commission (HRVC) and introduced a companion bill to the legislature. The bill directed the Commission to conduct a study of the river corridor—from the Verrazano Narrows in New York City to the boundary of the Adirondack Park—and report back to the legislature within a year with proposals for a program to protect its scenic, historic, and cultural resources. Rockefeller also announced a proposal for a billion-dollar Pure Waters Bond Act aimed at cleaning up sewage pollution in all the waterways of the state and spotlighting the Hudson as a primary area for attention. By the end of the year, both measures had been adopted. The bond issue, which required a public referendum, was approved by the voters by a four-to-one majority, a dramatic showing of public support.

While Rockefeller was talking up his two proposals, Ottinger scheduled local hearings on the Riverway bill. The hearings were very well attended. At a meeting in Nyack, New York, 350 people turned out in support of the Riverway bill. In the spirit of the times, folk singer Tom Glaser, a radio personality with station WQXR, sang a song to the assembled crowd. To the tune of "Glory, Glory Hallelujah," the lyrics were: "Glory, Glory Henry Hudson . . . soon your shores will just be mud son . . . let's replace your trees with electricities . . . it's a brave, brave, brave new world." Supreme Court Justice William O. Douglas sent a letter to the assembled citizens supporting the Riverway bill and urging them on. "Good luck to you," Douglas said, "The Hudson is a rare and unique treasure."[24]

A few months later, Rockefeller's Hudson River Valley Commission was holding hearings in preparation for delivering its report to the legislature. A star-studded cast had been assembled to form the Commission membership: former Governor Averell Harriman; Henry Heald, President of the Ford Foundation; Marian Sulzberger Heiskell of the *New York Times*; Frank Wells McCabe, President of the Albany-based National Commercial Bank and Trust Co.; Alan Simpson, President of Vassar College; author and news commentator Lowell Thomas; William H. Whyte of the American Conservation Association; and Thomas Watson, head of IBM, which had a major business presence in the Hudson River Valley. The Governor had installed his brother Laurance as Chairman.

The Hudson had not received such high-powered attention since the turn of the century. From March to December, the year 1965 was punctuated by hearing after hearing about the Hudson and Storm King.

It was at this time also that the U.S. Court of Appeals for the Second Circuit was reviewing the FPC license to Con Ed. On October 8, 1965, the court heard oral arguments on the Storm King case, with Con Ed reemphasizing the need for the pumped storage project as an emergency power supply, citing the economic value of the plant and noting how noneconomic values, such as scenery, would be mitigated. The FPC, whose grant of the license was the subject of the appeal, defended its action and also challenged the right of Scenic Hudson to sue. The FPC claimed that the organization did not have legal "standing" because it could claim no personal economic injury from the FPC action. "The pattern of administrative law would be seriously undermined if any interested person or group were permitted to undertake the role of representative of the public interest in the courts," the FPC wrote in its brief. Scenic Hudson repeated its case for protection of a natural heritage and challenged Con Ed's plans on the basis that other alternatives were available which could meet the region's power needs while also protecting the mountain. Its lawyers argued that the effects on the striped bass fishery had not been adequately evaluated, and that the FPC had failed to consider fishery and scenic impacts in the context of additional power plants planned by Central Hudson at Breakneck. In these arguments, Scenic Hudson maintained that the alternatives it had presented were feasible although not necessarily better, but that, in light of the extraordinary significance of the Highlands, it was appropriate that the FPC at least consider them. Instead, Scenic Hudson noted, the Commission had refused.

On November 9, 1965, shortly after the hearings were over, New York City suffered the worst power blackout in history. Con Ed's system experienced a massive failure. The utility was quick to point out that the pumped storage project was designed to avert such disasters. Scenic Hudson countered that the blackout was a result of Con Ed's mismanagement.

It was rumored that the Court of Appeals wrote its decision by candlelight in the dark hours of the blackout. On December 29, 1965, the Court announced its decision: it was a victory for Scenic Hudson. The Court sent the case back to the Federal Power Commission, requiring it to consider scenic and historic resources and alternatives. It was the first time the court had ever reversed an FPC license for a power plant. Lawyer Al Butzel recalls being so shocked that when a reporter from the *New York Times* called for comments, he had nothing to say.

The decision of the Court was a landmark in legal history. At the core of its decision was the following statement: "The Commission's renewed proceedings must include as a basic concern the preservation of natural beauty and of national historical shrines, keeping in mind that, in our affluent society, the cost of a project is only one of several factors to be considered."

The court also directed the FPC to actively investigate the issues raised by citizen groups, noting: "In this case, as in many others, the Commission has claimed to be the representative of the public interest. This role does not permit it to act as an umpire blandly calling balls and strikes for adversaries appearing before it; the right of the public must receive active and affirmative protection at the hands of the Commission."

With regard to the FPC's assertion that Scenic Hudson had no right to sue, the court stated that "In order to insure that the Federal Power Commission will adequately protect the public interest in the aesthetic, conservation, and recreational aspects of power development, those who by their activities and conduct have exhibited a special interest in such areas, must be held to be included in the class of 'aggrieved' parties."[25]

Con Ed appealed the decision to the Supreme Court, which refused to hear the case, letting the Court of Appeals decision stand. The decision established a major precedent for environmental protection by giving resource protection equal weight with economic values and by guaranteeing citizen groups the right to argue for the protection of environmental interests in court. The case gave birth to a whole new class of environmental groups organized to represent citizen interests in legal proceedings, among them the Natural Resources Defense Council and the Environmental Defense Fund. The case was the precedent for most citizen suits today. Over the next four years, the concept stated by the Court of Appeals would be refined by the federal courts and ultimately by Congress and numerous state governments, gradually clarifying the public interest in natural resources.

Following this logic, Congress adopted the National Environmental Policy Act (NEPA) in 1969, requiring an environmental impact review of all major projects proposed by or requiring approval from the federal government. Many states, including New York, followed suit with their own "little NEPAs" requiring environmental review of state projects. Thus, the principle established in the Storm King case soon became the law of the land.

The Court of Appeals, however, did not guarantee the protection of Storm King Mountain; it merely directed the FPC to reconsider. Nevertheless, for Con Ed, the picture did not look bright and would soon look even worse. The December 1965 court decision was followed in February 1966 by the report of the Hudson River Valley Commission, which contained the following recommendation about Con Ed's plans to build a power plant on Storm King Mountain:

> The Commission strongly believes scenic and conservation values must be given as much weight as the more measurable economic values and that we should not necessarily destroy one value to create another.
>
> The immediate case in point is the plan of Con Edison to build a pumped storage plant at Storm King Mountain. The Commission believes that scenic values are paramount here and that the plant should not be built if a feasible alternative can be found.[26]

This was no small statement, coming from a group of such influential leaders. It did nothing to make life easy for Governor Rockefeller whose brother and cousin were part of the Commission's leadership.

Still, the final decision rested with the FPC, which scheduled a new round of hearings for November 1966. At the court's direction, the FPC probed exhaustively into the issues raised by Con Ed's opponents: scenic concerns, alternate energy sources and fishery impacts. More than 5000 pages of testimony were presented by the time the hearings closed on May 23, 1967.

It was Scenic Hudson's task to prove that the Highlands and Storm King were scenic and historic, and that these values should determine the use of the land. The testimony before the FPC makes for fascinating reading. Among those who testified were Vincent Scully, professor of art history at Yale University, who compared Storm King to sacred shrines of Greece, and Anthony Wayne Smith, President of the National Parks Association, who suggested that the Highlands be added to the National Park system. A. Rudolf Wunderlich, a dealer in Hudson River School paintings for the Kennedy Galleries, testified to the importance of the Hudson River School in American art history and argued that the scenes the artists had painted should forever be preserved. Historian Carl Carmer recounted the story of the role played by the Highlands in the American Revolution. Richard Edes Harrison, a New York City cartographer, compared the geographic and scenic features of the Hudson Highlands gorge to other water gaps such as the Connecticut River at Mount Holyoke, the Potomac at Cumberland, the French Broad as it flows through the Smokies, and the Delaware Water Gap. Charles W. Eliot, 2nd, a professor emeritus of City and Regional Planning at Harvard University and a renowned landscape architect by profession, evaluated the impact of the proposed construction on the natural beauty of the Highlands.

Never was there such an outpouring of praise for the river. In many cases it was pure poetry. For those who were there to hear it, Scully's imagery was among the most powerful. The following is an excerpt from his testimony under questioning from attorneys:

> *Q.* Do you feel that the pumped storage project, as now proposed by Consolidated Edison, would damage the scenic features at Storm King Mountain?
> *A.* I do. I believe with Professor Eliot, who has described the impact in detail, that the Consolidated Edison proposal would seriously injure one of the most valuable and unusual natural formations and scenes in the United States—Storm King Mountain.
> *Q.* What is the basis for your conclusion?
> *A.* I leave aside the question of woodlands drowned in the enlarged reservoir, as well as that of power lines marching across the folded hills on the eastern side of the river. Those features would probably do damage enough.
>
> But Storm King is the central issue, and it is a mountain which should be left alone. It rises like a brown bear out of the river, a dome of living granite, swelling with animal power. It is not picturesque in the softer sense of the word, but awesome, a primitive embodiment of the energies of the earth. It makes the character of wild nature physically visible in monumental form. As such it strongly reminds me of some of the natural formations which mark sacred sites in Greece and signal the presence of the Gods; it preserves and embodies the most savage and untrammeled characteristics of the wild at the very threshold of New York. It can still make the city dweller emotionally aware of what he most needs to know: that nature still exists, with its own laws, rhythms, and powers, separate from human desires.[27]

The testimony of Brooks Atkinson, *New York Times* drama critic and a Pulitzer-prize winning journalist, encapsulated the history of the Highlands over three centuries, emphasizing the depth of America's cultural roots in the region:

> Every generation stands on the lofty pyramid of past experience and accomplishment. The grandeur of the Hudson River Gorge contains many things that give depth and significance to our life today—first the ancient geological formation that made a splendid river out of a wide arm of sea; then the modest, inquiring voyage of Henry Hudson in the autumn when the scent of wild grapes gave the air a spicy perfume and the Indians brought him gifts of corn; then the era of the American Revolution when some blundering, ill equipped American farmers managed to keep the down river British from joining the Albany British; next, the romantic era when sloops sailed through the gorge; finally, the happy period when Hudson River steamboats thundered through the gorge and carried thousands of city dwellers up and down river. Thus, the gorge represents some fine American traditions as well as the physical grandeur of the rock-walled passage.[28]

By this time, there was also considerable concern that the Storm King plant would suck in the eggs and juvenile fish of the striped bass. As Robert Boyle had testified in 1964, striped bass were known to spawn at the base of Storm King. As word of potential fishery impacts spread, powerful new allies joined Scenic Hudson's cause for the second round of hearings. Communities on Long Island and Connecticut, whose economic base relied heavily on bass fishing, registered their opposition to the power plant, as did boat owners' associations and sportsmen's clubs. The state legislature of Connecticut adopted a resolution opposing the New York power plant. Nassau County, on Long Island, intervened on the side of Scenic Hudson, as did more than forty other towns, counties, individuals, and conservation groups. Concern over these issues, together with the fish kills at Indian Point, led Boyle to form the Hudson River Fishermen's Association as a voice for sport and commercial fishermen. HRFA joined in the lawsuit and brought ecologist Ronald Dagon to the hearings to testify.

In an effort to avoid further controversy, Con Ed, before the hearings began, proposed further modifications in its plant design, offering to construct the entire powerhouse underground. This would raise the project cost to $169 million and constituted a substantial concession.

However, concerns about fishery impacts remained unresolved, and by this time it was clear that for many people any power plant at Storm King would be an industrial intrusion in an area which had become regarded as the symbol of wild, unsullied nature. Furthermore, the mountainside would still be cut way at its base, leaving a hole the size of two football fields with a stone wall 35 feet high behind it and concrete abutments at each level. While Con Ed tried to play down these features, however, they doubtless represented a visible encroachment.

A new view of the Highlands was emerging which tolerated no compromise. As conservationist Richard Pough stated in his testimony to the FPC:

> I have no doubt that the replanting, the painting and the other landscaping can cover some of the scars of the project site, allowing several years for the growth of the shrubs, vines and trees to be planted. But the ultimate effect of the project cannot be avoided. The integrity of the Mountain will have been destroyed. It must be remembered that Storm King is not merely landscape. Storm King is a particularly moving embodiment of the land itself. It is the fabric of natural forces—not merely picturesque. The intrusion of industry, however disguised, would necessarily commit the Mountain to a hostile use.[29]

Richard Harrison in his testimony also supported this view: "No matter how many bushes they plant on the dams, the hand of man will always be evident, and this is fatal to the idea of wilderness."[30] Storm King struck a primal chord in American hearts. Many people felt that the time had arrived when the Highlands should be inviolate. Furthermore, they feared that allowing a power plant at Storm King would open the door to further industrialization of the mountains. With Central Hudson waiting in the wings to develop its own plant, the threat seemed very real.

Con Ed's proposal to put its powerhouse underground was to have additional ill-fated effects on the utility. The utility's scheme not only failed to mollify Scenic Hudson and other opponents, but also suddenly raised the ire of New York City. The new underground powerhouse would require blasting in the vicinity of the aqueduct tunnel, an area known to be unstable; the Storm King water tunnel had been plagued in 1913 by "popping rock" and shifting granite. Since the aqueduct provided forty percent of New York City's water supply, Mayor John Lindsay was justifiably concerned. In 1968, the City decided to oppose the project and one of Lindsay's representatives testified at the hearings not only on the danger to New York's Water Supply but also on the importance of protecting the scenery.

While the FPC conducted its four years of hearings, members of Congress kept the Hudson on the political agenda, and Scenic Hudson kept Storm King in the news. On a cold November morning in 1967, Scenic Hudson chartered the tour boat "Miss Circle Line." The organization's annual fundraiser that year would be on the river itself, with a trip from Manhattan up to Storm King. Senator Robert Kennedy and Congressman Richard Ottinger were among the many luminaries scheduled to speak. The boat and the refreshments had been donated for the cause. Reporters and photographers from twenty-three news media boarded the boat in Manhattan along with 300 passengers. Television reporter Marya Mannes was one of those on board and gave this review of the event:

> I went upriver with Bobby Kennedy last Saturday.
>
> So did three hundred other people. That chilly morning we boarded the flagship of the sightseeing fleet, Miss Circle Line, and glided northward up the Hudson to look at Storm King Mountain.
>
> You may have guessed by now that we were a band of conservationists going to have a first-hand look at the great mass of prehistoric rock that Con Ed wants to dig and we think it mustn't . . .

> After speeches we went on deck to look at Storm King. The cold mist still shrouded the top of this magnificent guardian rock. But we were unmistakably in the presence of greatness.
>
> Here was the untempered soul and body of earth itself, part of the organic whole of nature from which man has immemoriably drawn his strength . . .
>
> Conservation is not a matter of loving nature. It is a matter of saving man.[31]

The Storm King issue now appeared in the national press on a regular basis, making news in faraway places. The Sunday edition of the *San Francisco Examiner and Chronicle* carried this headline on March 2, 1969: "Showdown Near in Fight to Harness Granite Hulk of Storm King." The *News-Herald* of Conneault, Ohio carried a story on March 5, 1969 titled "Conservationist vs. Engineer: Big Storm Brews over New York's Scenic Storm King Mountain." In Massachusetts, Harvard University found itself under attack in the student newspaper, the *Crimson*. As the owner and potential seller of Black Rock Forest, the site of the proposed plant's reservoir, it was accused in the *Crimson* of "environmental rape."[32]

Much of the publicity focused on the Revolutionary War significance of the Highlands. Although the most important battles took place at Forts Clinton and Montgomery at the southern gate to the Highlands, and the later events centered around West Point, in the public consciousness, Storm King bore the heroic mantle of the Revolution. The June 1967 issue of *American Heritage* carried a long article by Carl Carmer describing "how the battered survivors of a grim Revolutionary War battle hallowed this contested ground."[33]

Con Ed, faced with a media barrage, put its own public relations people to work, producing a full-color booklet titled "The Cornwall Hydroelectric Project: Power and Beauty for Tomorrow," which attempted to undo the damage done by the graphic artist in the 1962 annual report. The utility also put out a newsletter of favorable press clippings, while its lobbyists labeled powerplant opponents as "misinformed bird watchers, nature fakers, land grabbers and militant adversaries of progress."[34] Another Con Ed p.r. piece was a thirty-page booklet titled "The Truth vs. Twenty Unfounded Charges." Despite such efforts, the media, for the most part, threw their backing to Scenic Hudson. Con Ed had gained the reputation, as *Fortune* magazine put it, of the "Company you love to hate."[35]

Eventually thousands of people from nearly all American states and several foreign countries flocked to the cause. At the height of its activity, Scenic Hudson had 22,000 contributors from 48 states and 14 foreign countries. Chairwoman Frances S. Reese took particular pleasure in the donations she received from Con Ed stockholders who gave their dividend checks to help fund the lawsuits. Many prominent people backed Scenic Hudson publicly, including James Cagney, Henry Morgan, and Aaron Copland. Pete Seeger tried to get Scenic Hudson to support his idea of building a replica of a Hudson River sloop, which would sail the waters of the Hudson promoting the idea of cleaning it up. Scenic Hudson, already stretched thin by supporting its director, office, and

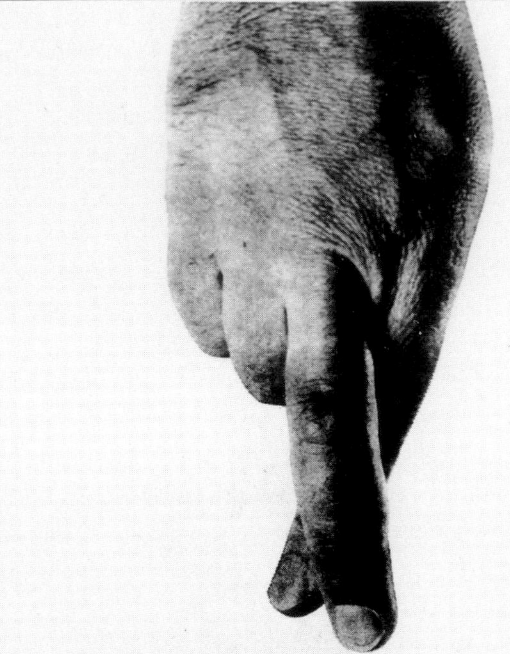

When Con Edison starts dynamiting Storm King Mountain, keep your fingers crossed.

40% of your water supply may go down the drain.

As early as this November, according to a recent Con Edison announcement, the utility could close an electric contact and trigger an explosive charge planted in Storm King Mountain on the Hudson River near Cornwall. It will be the first of thousands of charges necessary to dislodge 580,000 tons of rock. When the debris is cleared away, three underground chambers will be left, the largest of which could accommodate a 15-story building and will be the site of a huge powerhouse for Con Edison's proposed Storm King Pumped Storage Generating Plant.

But there's a hitch.

Just 140 feet away from a corner of the powerhouse chamber is a tunnel–part of the Catskill Aqueduct–that carries 40% of New York City's water supply (and up to 100% of the water supply of a number of Westchester communities).

The first attempt to construct this tunnel, in 1913, failed because surrounding rock formation could not support even a 20-foot boring. A new, smaller tunnel was dug. Again, unstable rock was encountered but the tunnel has held. So far.

We Need All the Luck We Can Get

But what happens to an old tunnel–which New York City engineers have called "fragile"–when a dynamite blast is set off just a stone's throw away. If we're all lucky, nothing.

What happens to that tunnel over a period of months when thousands of explosive charges, representing tons of dynamite, are triggered just a stone's throw away. If our luck holds, nothing.

And what happens to the surrounding formation when it's undercut by the removal of more than half-a-million tons of rock? If we all keep our fingers crossed, maybe nothing.

And if we're unlucky?

The aqueduct breaks.

And New York City–its people, its industry, its schools, its hospitals–would have to get along with 40% less water. Things would be even tougher for hundreds of thousands of Westchester residents, for whom the Catskill Aqueduct is the sole source of water.

There are some people who think we shouldn't have to trust to luck to keep our water supply secure. The Scenic Hudson Preservation Conference, for example. Scenic Hudson has been fighting the Storm King project ever since Con Edison first announced plans for it ten years ago. The struggle has been waged on many fronts–environmental, scenic, economic–but perhaps no issue is likely to be as critical to the immediate well-being of millions of people as the risk of losing a major portion of their water supply.

City Calls Possibilities "Catastrophic"

The City of New York, balancing the need for more power against the need to safeguard its water supply, opposes the project. As the Corporation Counsel for the City of New York argued in a brief before the Appellate Division of the New York Supreme Court:

> "The City's engineers fear that such a massive excavation so close to the Aqueduct tunnel will set off stress changes in the rock, thereby causing the Aqueduct to break. *Such a break could be catastrophic.*"

Con Edison has characterized the risk as "remote." The Federal Power Commission, in granting Con Ed license to proceed with the project, apparently agrees. But in the appellate review, which refused to reverse the FPC ruling, a dissenting judge wondered whether *any* risk of this magnitude was worth contemplating. He wrote:

> "There is a world of difference between no danger and 'remote' danger. If a danger is 'remote' the degree of 'remoteness' assumes importance in proportion to the magnitude of the danger. Here the danger is obviously great, and there is no finding as to the degree of remoteness."

Originally, when Con Edison proposed to locate the plant above ground, a consulting geologic engineer testifying on behalf of the utility saw no danger to the aqueduct. But when the courts rejected an above ground installation, and Con Edison proposed the present underground site, this same geologic engineer was moved to the opposite conclusion, and testified on behalf of New York City as follows:

> "The risk of failure of the aqueduct cannot be regarded as imminent but *it represents a definite hazard* . . . The risk might be taken . . . if only money were involved; however, a failure of this water supply might jeopardize the lives and welfare of millions of persons in the City and the upstate communities served by the Catskill Aqueduct."

In other words, when they start dynamiting Storm King Mountain, keep your fingers crossed.

What Can You Do?

Fortunately, Con Edison isn't home free yet. The City is still battling in court to block the project and protect its vital water supply. Your support is needed.

Uncross your fingers long enough to write:

Governor Rockefeller
The Executive Chamber
Capitol
Albany, New York 12224

Mayor Lindsay
Office of the Mayor
City Hall
New York, N.Y. 10007

Martin Lang
Commissioner of Water Resources
2454 Municipal Bldg.
New York, N.Y. 10007

Charles F. Luce, Chairman
Con Edison
4 Irving Place
New York, N.Y. 10010

Write your Congressman, too.

Tell them the risk is too great for Con Edison to dynamite Storm King.

Write Scenic Hudson Preservation Conference. We'll tell you about committees you can join, petitions you can start, money you can contribute.

Water is one of the few things New Yorkers and Westchesterites haven't had to worry about lately. Con Edison may take care of that, too.

Scenic Hudson Preservation Conference

545 Madison Avenue, New York, N.Y. 10022

This space paid for by friends of Scenic Hudson Preservation Conference.

Scenic Hudson full page ad, dated October 10, 1973, warning about the potential effects of the Storm King project on the New York City aqueduct, Storm King tunnel. Courtesy of Scenic Hudson, Poughkeepsie, N.Y.

lawyers, refused, so Seeger did it on his own, forming the organization which is now the Hudson River Sloop Clearwater. The *Clearwater* sloop, constructed during the years of the FPC hearings, was launched in 1969.

Throughout this time the politics of the Storm King issue continued to escalate. As early as 1965, the House Subcommittee on Fisheries and Wildlife had held hearings on the Hudson River striped bass fishery. Department of Interior staffer James McBroom had testified at those hearings that the Storm King project "could have a serious effect on the Hudson River striped bass fishery and the dependent fisheries around Long Island and offshore."[36] Secretary of Interior Stewart Udall became a vocal opponent of the project and began to attack Governor Rockefeller.

Conservationists made numerous attempts to involve President Johnson, who had made the preservation of natural beauty a cornerstone of his Presidency, but to no avail. When Johnson made his historic speech to the Congress on the subject in February 1965, he outlined a program of action for the cities and the countryside, and listed twelve sites to be added to the national system of parks, seashores, and recreation areas; but he made no mention of Storm King.[37] His White House Conference on Natural Beauty held in May 1965, also passed the subject by, to the dismay of preservationists.

When Lady Bird Johnson toured the Boscobel mansion in Cold Spring on May 17, 1968, Scenic Hudson sought to engage her concern. At a dinner reception, Ander Saunders, a Scenic Hudson founder, lobbied her by offering a toast: "To a gracious lady who has come to see our Hudson Highlands—long may the Highlands from Storm King to Dunderberg and from Mount Beacon to Anthony's Nose remain strong in their God-given dignity and beauty, unsullied, for the inspiration and recreation of man." Lady Bird graciously accepted, but the visit failed to generate any involvement by the President, whose silence on this high-profile issue was baffling.

Johnson did sign the Riverway bill, however. The bill, later referred to as the Interstate Compact—or more often, the Hudson River Compact—became law in September 1966. It had been substantially revised and required ratification by the states of New York and New Jersey within three years. The Governor, reluctant as ever to give any new authority to the federal government or to New Jersey, jealously guarded his jurisdiction over the Hudson. He disputed the ratio of membership among New Yorkers and representatives of other states and the federal government, and succeeded in tactics of stall and delay. In 1970, New York state had still not ratified the Compact, and the deadline was extended for another four years and finally lapsed in 1974. The mood, by then, had changed and the Riverway bill was never revived.

One reason was the increasing vigor of the Hudson River Valley Commission. After presenting the results of its study to the legislature in 1966, it was reconstituted by the Governor and the legislature as a permanent Commission with power to review projects visible from the river within a mile-wide corridor. The creation of the Commission was an effective preemptive strike. Governor

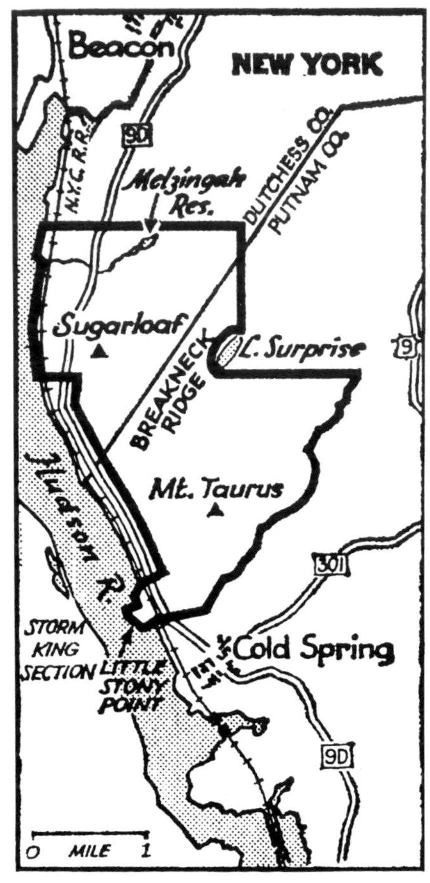

Location of Hudson Highlands state park. Courtesy of *The New York Times.*

Rockefeller stole Ottinger's thunder, and as a result, the federal compact law withered on the vine.

The Hudson River Valley Commission had no power to stop a project but it had the power to delay it through public hearings. The HRVC also had a large staff of landscape architects and other professionals to advise developers on alternate site designs which would decrease the impact of a project on the Hudson's shoreline scenery. The HRVC influenced many projects, from housing units to shopping malls and power plants. One of its more significant contributions was the alteration of the design for an oil-fired generating plant to be built by Central Hudson at Danskamer Point in Newburgh just north of the Highlands. Central Hudson's president was very concerned that the design of the Roseton plant fit into the landscape. He approached Scenic Hudson for advice and ultimately adopted a design suggested by HRVC. The "Roseton" facility, as finally approved, was constructed so that it would not rise above the ridge behind it. From the north, looking south, it is invisible except for a smokestack. From the south, its oil tanks are screened by an earthen berm, which also limits the view of the plant itself. In the growth years of the 1960s HRVC reviewed and redesigned countless other projects.

In 1966, the Hudson River Valley Commission conducted a study of the Hudson Highlands. Its principal recommendation was the creation of a park on the east bank of the river on Breakneck Ridge. This was backed by a similar recommendation by the State Council of Parks. In 1967, the Georgia Pacific Company announced that it had purchased Little Stony Point, just below Breakneck Ridge from the Hudson River Stone Company, with the intention of constructing a wallboard manufacturing plant, a proposal which raised a great hue and cry from citizen groups and Congressman Ottinger.

The company needed a site with access to the Hudson to bring in gypsum by ship from South America. Little Stony Point was part of the Hudson Highlands park envisioned by the Valley Commission. The Commission (with help from William H. Osborn) involved Governor Rockefeller in persuading the company to relocate. The Governor interceded, finding a location for Georgia Pacific's gypsum plant in nearby Verplanck.

Meanwhile Central Hudson, which had kept its pumped storage project plans on ice, was approached by Laurance S. Rockefeller, Chairman of the HRVC, with a proposal to buy its lands for park purposes. Rockefeller offered to purchase the land on behalf of New York State through the Jackson Hole Preserve, a family foundation devoted to conservation. Central Hudson agreed to sell the 670 acre Breakneck property to the state for the same amount of money it had paid for it, \$827,789, stating that: "the use of the property for park and conservation purposes would be in the best interest of the company and the public which it serves."[38] Ultimately, Rockefeller assembled a plan for a 2500-acre park including Breakneck, Mount Taurus, Bannerman's Island and Little Stony Point. The cost was \$3 million, of which half was paid by Rockefeller family foundations.

With great fanfare, Governor Nelson Rockefeller held a ceremony on May

23, 1970 dedicating the land as the new Hudson Highlands State Park. Little Stony Point, formerly a quarry, is now a popular beach for swimming and boating, which affords some of the best views of the Storm King gorge. Among other things, Governor Rockefeller clearly hoped that the purchase of Breakneck would be the political compromise that would allow the Storm King project to move forward.

A few months later, on August 19, 1970, the Federal Power Commission concluded four years of review of the Storm King project. In a lengthy decision, the FPC set forth its findings that:

> The Hudson Highlands is an area of great natural beauty, having historical and artistic significance and exceptional recreational opportunities. . . . The construction of the project . . . would not adversely affect, to any significant extent, the natural beauty, the historical significance, or the recreational opportunities of the area in which such a plant would be situated. . . .
>
> The project will not adversely affect the fish resources of the Hudson River provided adequate protective facilities are installed. . . .
>
> Cornwall makes the best use of available resources to meet the require-

Breakneck Mountain. Copyright © 1990 by Robert Beckhard.

> ments for electric energy with the minimum adverse impact on the environment. . . .
>
> This license is issued to Con Ed.[39]

Scenic Hudson and the Hudson River Fishermen's Association appealed the decision, but this time the license was upheld by the Court of Appeals. Justice Oakes dissented, stating that

> the commission's finding overlooks the fact that we are considering here a power station which above ground will consist of a concrete tailrace with abutments 32 feet high and 685 feet long, cutting back existing shore line from 195 to 260 feet, exclusive of any access road. . . . the mountain may 'swallow' the project, but the concrete tailrace and abutments, as long as a good-sized football stadium—over an eighth of a mile—and three stories high, will surely stick in its craw.[40]

The Supreme Court again refused to hear the case, allowing the license to stand.

It was a major victory for Con Ed and a serious setback for Scenic Hudson and the Fishermen's Association. However, the legal challenges were not yet over. Three more years were spent in New York State courtrooms challenging water quality permits granted to Con Ed by the Department of Environmental Conservation. Then, in December 1973, the results of new fishery studies and evidence of possible impacts were presented to the U.S. Court of Appeals by the Hudson River Fishermen's Association along with Scenic Hudson. The Fishermen's Association argued that the Con Ed studies had failed to take into account that the river was tidal. In July 1974, the Court of Appeals ordered further hearings on the fishery issues.[41] While the utility conducted several more years of court-ordered striped bass studies, the Storm King project was put on hold.

In the course of these hearings and studies, Con Ed had acquired new leadership. Waring and Forbes had both left the company, and Charles Luce came on as the new Chairman of the board. Luce strongly advocated the Storm King project but began to recognize the natural environment as a public concern which utilities must consider.

Luce began to speak and write on the issue as it was faced not only by Con Ed but also by all American utilites. Speaking before the New York City Bar Association, he said:

> The incompatibility . . . between protection of the environment and production of electricity is not absolute. Accommodations can be made which recognize the validity of both social objectives. . . . It is not correct to say . . . that only a "handful of people" are expressing this concern by opposing power plants. Every Gallup poll taken on the subject in the past few years shows that concern about air and water pollution ranks very high in any list of the citizenry's concerns about national problems. . . . I think it is accurate to say that the senior management of the utility industry now recognizes its responsibility to supply the electric energy needed by society in ways that will not unduly damage the natural environment.[42]

Luce began to deal with the project in a more practical light. From 1969 to 1971, to meet a power crisis in New York City, he quickly installed generating capacity in the form of 2.5 million gas turbines, most of them located in Brooklyn and Queens. Years later, after several more rounds of hearings, lawsuits and appeals, Luce began to consider the possibility of a mediated settlement. Con Ed, by this time, had become embroiled not only in battles over the Storm King plant but also in lawsuits over the fish kills at Indian Point. Luce saw the possibility of a solution which could produce benefits to both sides. Through an intermediary he approached Russel Train, former chief of the Environmental Protection Agency for his assessment. Train eventually was selected to settle a variety of Hudson River power issues as a mediator for eleven interested parties including the federal government, two state agencies, three environmental groups, and five utilities.

These issues centered around the protection of the fishery and the scenery while providing for long-term power generation by the utilities. In December 1980, after seventeen years of bitter controversy, a settlement was announced which ended the Storm King battle forever. The settlement provided for the pumped-storage project to be abandoned and the land donated by Con Ed to the PIPC and to the Town of Cornwall for a park. It also required that the utilities and the state Power Authority take measures to reduce the fishery impacts from their cooling systems at the Indian Point nuclear power plants. In return for these concessions, Con Ed and other utilities operating on the Hudson River won a reprieve from building expensive cooling towers on their riverfront power plants while technologies and operating procedures to reduce fish kills could be explored. In addition, the settlement required the utilites to provide a $12 million endowment for a new "Hudson River Foundation" to fund independent research on the Hudson River ecosystem to provide a base of information needed to manage the resource for its long-term conservation and development. The settlement was hailed as a "peace treaty on the Hudson" and a new model for resolving disputes by mediation instead of lawsuits.

Looking back at the records of seventeen years of legal battles, it seems that the fight to save Storm King revolved around a host of issues including scenery, history, water supply, fisheries, energy production, and the obligations of government agencies to balance and protect all of these public purposes. In the press it was alternately played up as a David and Goliath struggle of a small band of conservationists against a mighty utility and a soulless bureaucracy; as a thinly disguised ruse of rich men and women to protect the view from their backyards; as the struggle of towns and citizens to expand their tax base against outsiders seeking to preserve valueless scenery; and as an excuse to fight "the company you love to hate." The Storm King battle was all of these things and more. Ultimately, it was the story of people rallying to protect their spiritual connection with the land. The Highlands had become an inviolable, sacred landscape.

Sacred lands have a power beyond mere physical presence. Such lands have acquired layers of meaning, preserved with undiminished intensity through generations. Native Americans have died in great numbers to preserve their sacred lands. To enter Jerusalem is the dream of Jews all over the world. Thousands of Muslims each year travel to Mecca. To those who will never see the holy land, there is the knowledge that it is there.

It is a matter of history that certain landscapes have the power and meaning of holy lands. The notion of our landscape as symbolic of our national identity was expressed by artists and writers in the nineteenth century. The idea of the landscape's spiritual value became the driving force behind the preservation of the national parks in the west and the creation of a "forever wild" forest preserve in New York. Storm King and the northern gate to the Highlands had become such a place.

Susannah Hickling Lewis Willard, Hudson From Cozzen's West Point, 1862 watercolor. Courtesy of the Society for the Preservation of New England Antiquities, Boston, Mass.

Epilogue. The Power of Place

If events had been allowed to take their course, two power plants, a prison, a gypsum plant, and tract housing would occupy lands which have since become parks and historic districts. Instead, a different type of development has occurred in the Highlands, beginning at a time when the idea of harmony with nature drew strength and approval from the most influential members of society and maintained by those who came after. All types of people joined in this process—business people, artists, the military, industrialists, average citizens, and today the Highlands have what most places have lost: a sense of place and identity.

It was Andrew Jackson Downing who showed us that we can develop the land and build our structures on it in a way that complements the landscape. For him, the issue was not so much whether the land would be developed but how. He made us aware that our choices reflect our relation to the land and are ultimately a statement of ourselves and our identity. As we look back over the history of the region, we see that each generation approached the land with its own context and set of values, but each respected what came before, adding something unique and contemporary but managing to hang onto a kernel of the Highlands' essential character.

This happened in the Highlands because of geologic history, when ancient processes molded a landscape of mountains and river that proved to be emo-

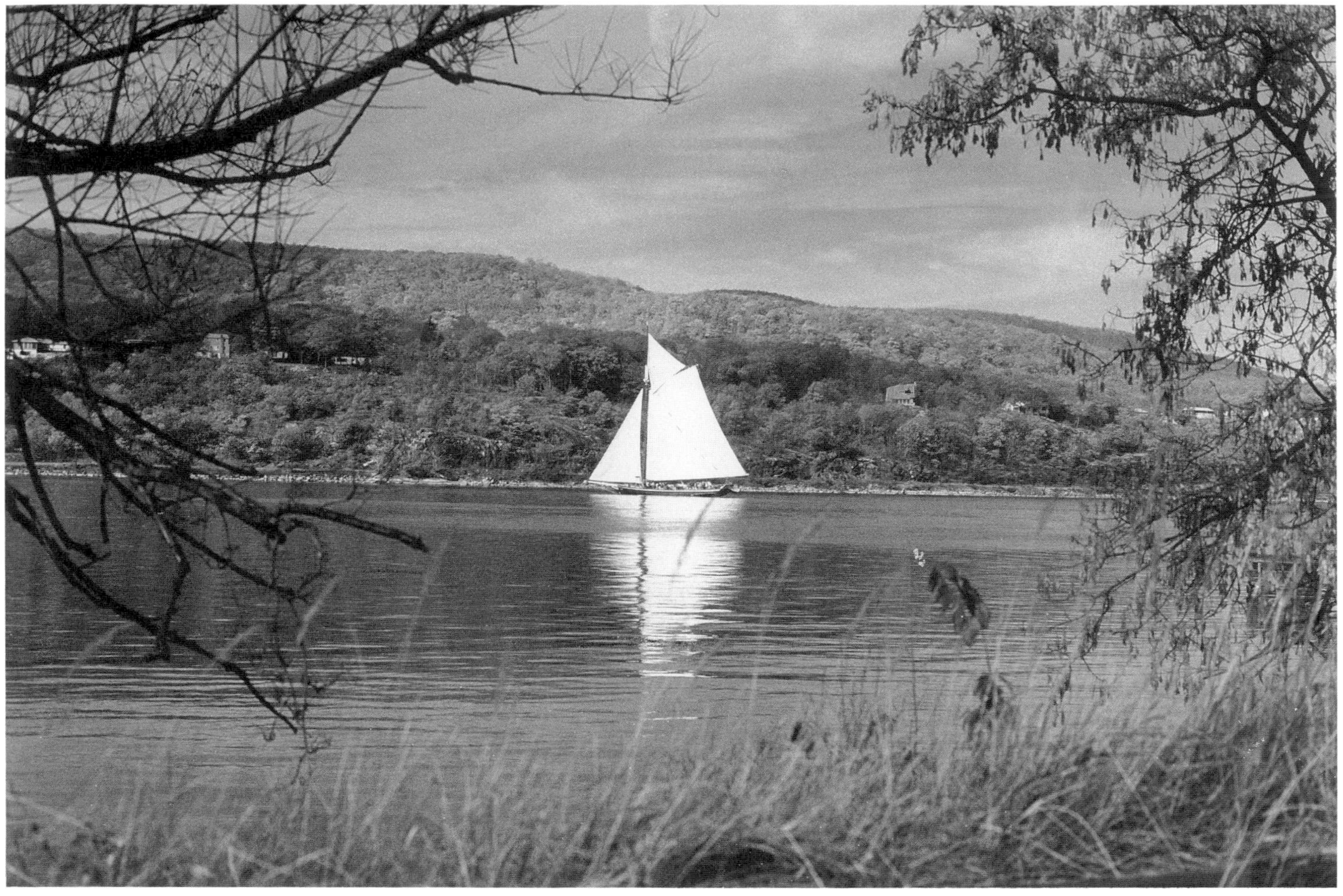

The Sloop "Clearwater" Copyright © 1990 by Robert Beckhard.

tionally evocative. While the response has been different over time, it has always been sufficient to wake us from the torpor of our daily lives, to reconnect us with our place on this earth and to act upon this knowledge.

Yet elsewhere, we find ourselves on a course which seems to have lost sight of this. The importance of the Highlands is that they remind us there is an alternative—one that involves conscious choices and the willingness to tackle the laws and politics which lead us down a path we may not want to go, until one day we get there and look back, wondering how it all happened. The Highlands also gave us the tools to exercise our rights as citizens in influencing the choice of directions.

The fight to save Storm King drew to a close in December 1980, but efforts to preserve the Highlands continue. In a long-established tradition, much of the conservation work has been carried on by private citizens. In the early 1980s, through the efforts of Scenic Hudson, 342 historic structures were added to the State and National Registers of Historic Places, including Revolutionary-war archaeological sites, colonial-era homes, great estates, WPA-built park lodges and three great monuments of engineering—the Storm King highway, the Bear Mountain Bridge, and the Storm King tunnel of the Catskill aqueduct. Similarly, in 1981, the federal government designated Iona Island and its tidal marsh, now a part of Bear Mountain State Park, a National Estuarine Research Reserve, to be managed for research and educational programs. It has also been listed as a National Natural Landmark.

Lila Acheson Wallace, who did so much to preserve Boscobel at its new location in the Highlands, endowed a foundation which, after her death, has continued to protect hundreds of acres of scenery. Most of the land preserved by the Wallace Fund remains in private ownership, protected by conservation easements in the manner begun in the 1930s by the Hudson River Conservation Society. A local land trust has also been organized by citizens of Philipstown.

In 1989, philanthropist William T. Golden organized a consortium of fifteen scientific and educational institutions to purchase the 3600 acre Black Rock Forest from Harvard University, which was no longer interested in using it as an experimental forest. The land, which in 1949 had been donated to Harvard with an endowment by Dr. Ernest Stillman (son of banker/financier James Stillman) had once been eyed by Con Edison for its Storm King power plant. It will now be preserved in perpetuity and used for ecological research and recreation. In like fashion, Frank and Anne Cabot are creating a forty-acre hilltop garden in Cold Spring, which they intend to endow and make public for the study of horticulture.

Private land conservation efforts have been matched by continuing initiatives by the state of New York. Borrowing from techniques established in the 1930s and revisited in the 1960s, the state legislature in 1978, under the sponsorship of Assemblyman Maurice Hinchey and Senator Jay Rolison, passed a Hudson River Study Act in 1978. The legislation called on the Department of Environmental Conservation (DEC) to identify scenic, ecological, and recreational resources in the river valley worthy of preservation. More recently, the legislature adopted new legislation directing DEC to develop a comprehensive plan for the management of the estuary, its fisheries, and its aquatic life. A state Greenway Council is also studying ways to preserve and connect parks, preserves, and historic sites.

These programs have led to the acquisition of important riverfront properties and the establishement of new state policies and programs. Denning Point in Beacon, a seventy-acre peninsula overlooking the Highlands, was added to the Hudson Highlands state park in 1988. Kowawese, also known historically as Plum Point, where revolutionary war chevaux-de-frises stretched across the river to Pollepel Island, was purchased in 1987 and will now be operated as a scenic preserve. This acquisition implements the idea advocated by N.P. Willis more than a century ago, when he wrote in his letters to the newspaper from "Idylwild": "Here should be a public park." The state has also protected the land around "Castle Rock" and the Constitution Island Marsh which is now managed by the National Audubon Society. As with the move to protect the Palisades,

Tide marsh at Iona Island, Bear Mountain State Park. The island and marsh have been designated as a National Natural Landmark and a National Estuarine Research Reserve because of their outstanding ecological value. Copyright © Steve Stanne/Clearwater.

such purchases have often been leveraged by private individuals offering to match state dollars in order to guarantee that these special places will be preserved forever as part of our natural heritage.

One of today's most pressing environmental concerns in the Highlands is the presence of cadmium, a toxic heavy metal present in Foundry Cove in Cold Spring. Located near the old iron foundry operated by Gouverneur Kemble, the cadmium deposit was dumped by a federally subsidized battery company operating in the 1950s and 1960s. Now a federally-designated Superfund site, slated for cleanup of toxic wastes, the cadmium dump has contaminated the sediments and aquatic life of the area. Blue-claw crabs, which migrate up and down the estuary, contain significant levels of cadmium in their hepatopancreas—the yellow mustard—causing the state health department to caution against eating this part of the crab. Pollution of the Hudson with PCB's has closed the striped bass commercial fishery, and the Hudson River Sloop Clearwater organization has been lobbying the state and federal government to remove the deposits of PCB-contaminated sediments. To date efforts by New York State have been unsuccessful due to lawsuits over the proposed methods and disposal sites, and the federal government has just recently decided to investigate the PCB hot spots as a potential Superfund project.

Another major concern is the expansion of New York City's water supply using fresh water drawn from the Hudson River north of the Highlands, where salinity is negligible. Issues such as movement of the estuary's salt-front and the effects of water withdrawal on spawning striped bass have been raised in hearings by Scenic Hudson, Clearwater, the Hudson River Fishermen's Association, and others; they will be the subject of ongoing public debate in the coming years. Meanwhile, some lakes in the Highlands have acidified—most likely the result of acidic soils in combination with air pollution from Midwestern power plants.

The Hudson River Foundation, created out of the Storm King settlement, with an endowment of what is now valued at some $25 million has supported scientific research on these issues. The foundation hopes over the long term to develop an understanding of the estuary's ecosystem which can be used to guide the management of the Hudson's natural resources.

This is the continuing story of the Highlands. It shows how a few individuals, when moved to action, can inspire thousands to follow, creating new institutions, new policies, new precedents. It reminds us that our relationship to nature can be observed and made a continuing part of our lives, and that this is, in fact, primarily a question of how we define ourselves—who we think we are.

The new environmental issues will be global ones and will demand that we adopt the values which have been tested in the Highlands—that we learn to treat the earth with the knowledge that it nourishes our bodies and spirits. We must be prepared to move ourselves and the governments of our own and other nations in this new, yet old, direction. The story of the Highlands has proved that it can be done.

Notes

1. *World's End*

1. Limburg, *The Hudson River Ecosystem*, p. 8 citing Schureman P., *Tides and Currents in Hudson River*, U.S. Coast and Geodetic Survey, Spec. Publ. 180. The U.S. Department of Commerce, National Oceanic and Atmospheric Administration, National Ocean Service, Nautical chart #12343, catalog no. 1, Panel F,G, "Hudson River, New York to Wappinger Creek" shows a depth of 175 feet. Other sources have noted that the Hudson here is 202 feet below sea level. All concur that the river is deepest as it rounds Gees Point at World's End.

2. Carroll, "Journal of Charles Carroll of Carrollton"; Van Zandt, *Chronicles of the Hudson* p. 81.

3. Wilkie, *The Illustrated Hudson River Pilot*, pp. 96, 98.

4. Robert Juet, "Juet's Journal," in Van Zandt, *Chronicles of the Hudson*, p. 14.

5. Van Zandt, *Chronicles of the Hudson*, p. 34, in "The Journal of Jasper Dankers" refers to Potlepel Eylandt dated 1680.

6. Hunt, *Letters About the Hudson and Its Vicinity Written in 1835 & 1836*, p. 24–25.

7. Irving, *Bracebridge Hall*, p. 246.

8. Carroll, "Journal of Charles Carroll," 81.

9. The discussion of geology and natural history of the Highlands is drawn from NYS Museum and Science Service, *Geology of New York*, pp. 5–15, Wykoff, *Rock Scenery*, pp. 17, 67, Sanders, *Principles of Physical Geology*, p. 451, Thompson, "Hudson Gorge in the Highlands," p. 1843.

10. Boyle, *The Hudson River*, p. 125.

11. Irving, *Bracebridge Hall*, pp. 267–269.

2. General Washington's Stronghold

1. Fitzpatrick, ed., *The Writings of George Washington*, p. 138.

2. Ford, *Journals of the Continental Congress*, 2:57.

3. Calihan and Larsen, "The Hudson Highlands," p. 4 quoting from Hornor, "The Obstruction of the Hudson River During the Revolution," *The American Collector* (1926) 3:436.

4. Clinton, *The American Rebellion*, 1:11–12.

5. Ford, *Journals of the Continental Congress*, 2:59–60.

6. Force, *American Archives*, Fourth Series, 2:1291, 1295.

7. Ibid. 3:1294.

8. Ibid. 3:1657–58.

9. Palmer, *The River and the Rock*, p. 44, quoting *Journals of Congress*, January 5, 1776.

10. The letters of July 12 and July 20 may be found in Fitzpatrick, *Writings of George Washington*, 5:266, 313.

11. Stedman, *History of the American War*, p. 361.

12. Ibid., *History of the American War*, pp. 405–406.

13. Dwight, *Travels in New England and New York*, pp. 305, 302–3.

14. Van Zandt, *Chronicles of the Hudson*, p.100 quoting Chastellux *Travels in North America, in the Years 1780–81–82*.

15. Palmer, *The River and the Rock*, p.170, quoting Thacher's *Military Journal of the Revolution*, 1862.

16. Moore, *Diary of the American Revolution 1775–1781*, p. 445.

17. Fitzpatrick, *Writings of George Washington*, 20:213.

18. Van Zandt, *Chronicles of the Hudson*, p. 102 quoting Chastellux *Travels in North America, in the Years 1780–81–82*.

19. Calihan and Larsen, "The Hudson Highlands," p. 51, quoting Maude's *Visit to the Falls of Niagara in 1800*, London: Longman, Rees, Orme, Brown and Green, 1826, p. 7, Journal for Monday June 23, 1800.

3. West Point: First Stop on the American Tour

1. Sutcliffe, *Robert Fulton and the "Clermont"*, pp. 207, 234.

2. Huth, *Nature and the American*, p. 72 quoting Clinton's *Letters on the Natural History and Internal Resources of the State of New York* pp. 10, 15; 1822.

3. Marryat, *A Diary in America*, p. 40.

4. Milbert, *Picturesque Itinerary*, p. 41; de Montule, *Travels in America*, Letter 23, p. 177.

5. T.S. Cummings *Historic Annals of the National Academy of Design*, p. 12 quoting Governor DeWitt Clinton's 1816 opening address at the first exhibition held by the National Academy of Fine Arts, a forerunner of the National Academy of Design.

6. Dickens, *American Notes*, pp. 221–221.

7. Huth, *Nature and the American*, p. 74.

8. Buckingham, *America, Historical, Statistic, and Descriptive*, pp. 237–244.

9. D'Auermont, *Society and Manners in America in a Series of Letters from that Country to a Friend in England By an Englishwoman*, (Letter ix, Albany, July 1819).

10. Van Zandt, *Chronicles of the Hudson*, pp. 211–212, quoting Martineau, *Retrospect of Western Travel* (London, 1838) 1:43–63.

11. Buckingham, *America, Historical, Statistic, and Descriptive*, p. 246).

12. quoted in Calihan and Larsen, "The Hudson Highlands," p. 59, who also make mention of the guidebooks I have cited above.

13. Colyer, *Sketches of the North River*, p. 58.

14. Willis, *American Scenery*, 1:v; Nash, *Wilderness and the American Mind*, p. 68, quoting *Prose of Freneau*, p. 228; Nash, *Wilderness and the American Mind*, p. 74, quoting Lanman, *Summer in the Wilderness*, p. 171; Byrant, quoted in Huth, *Nature and the American*, p. 36 no citation.

15. Leslie, *Graham's Magazine*, pp. 207–209, 290–295.

16. Forman, *West Point*, p. 94.

17. Leslie, *Graham's Magazine*, p. 294.

18. Ambrose, *Duty, Honor, Country*, p. 150.

19. Forman, *West Point*, pp. 99, 102–103, 91.

20. Poe, *The Best Known Works of Edgar Allan Poe*, pp. 551–552.

21. Gray, "The Architectural Development of West Point," p. 4–6.

22. Van Zandt, *Chronicles of the Hudson*, p. 200 quoting Frances Kemble Butler, *Journal* (1835).

4. *Landscape and Dreamscape*

1. The material on Cole is from Noble, *The Life and Works of Thomas Cole*, pp. 34–39, except for the reference to Dunlap's letter, which is on p. xviii.

2. Howat, *The Hudson River and Its Painters*, p. 27.

3. Van Zandt, *Chronicles of the Hudson*, p. 199 quoting Kemble *Journal of A Residence in America* (1835).

4. O'Brien, *American Sublime*, p. 124.

5. McShine, *The Natural Paradise*, p. 62, quoting Cole "Essay on American Scenery" (1835) and p. 70, quoting Durand, "Letters on Landscape Painting" (1855).

6. Noble, *The Life and Works of Thomas Cole*, p. 35.

7. The Alice and Hamilton Fish Library in Garrison, New York has started a comprehensive reference collection of the work of the Hudson River School. It now contains more than 500 photographic slides of paintings of Hudson River scenes, and well over 100 of the Highlands.

8. Irving, *A History of New York from the Beginning of the World to the End of the Dutch Dynasty by Diedrich Knickerbocker*, pp. 4, 386–387.

9. Bruce, *The Hudson*, p. 9.

10. Adams, *The Hudson River in Literature*, p. 28, quoting James Kirke Paulding's *New Mirror for Travellers; and Guide to the Springs* (1828).

11. Cooper, *The Spy*, p. 315.

12. Callow, *Kindred Spirits: Knickerbocker Writers and American Artists 1807–1855*, p. 119.

13. Huth, *Nature and the American: Three Centuries of Changing Attitudes*, p. 31; Bryant, *Thanatopsis*, p. 23.

14. Smith, *Edinburgh Review* (January 1820) 33:79.

15. Huth, *Nature and the American*, p. 30

16. Ibid. p. 49–50, quoting Goethe, "Die Vereinigten Staaten"

17. McShine, *The Natural Paradise*, p. 61.

18. Irving, *Crayon Miscellany*, p. 136, visit with Scott at Abbotsford in 1817.

19. Bruce, *The Hudson*, pp. 128–129.

20. Drake, *The Culprit Fay*, p. 13.

21. Willis, *Outdoors at Idlewild*, p. 188; compiled from Willis' letters to the *Home Journal*.

22. Mabie, *The Writers of Knickerbocker New York*, pp. 31–36. The quotes are on p. 35.

23. Howat, *The Hudson River and Its Painters*, p. 35.

24. Callow, *Kindred Spirits: Knickerbocker Writers and American Artists 1807–1855*, pp. 19–28; 169–170.

25. Nash, *Wilderness and the American Mind*, p. 71. Magoon was a minister who made a notable collection of Hudson River School paintings which he sold to Vassar College, where they continue to form the core collection.

26. Willis, *American Scenery*, p. 8.

27. Novotny, "Research Reports," pp. 1–4.

28. Melville, *Moby Dick*, chapter 41, p. 187.

29. Willis, *Outdoors at Idlewild*, p. 28.

30. Dictionary of American Biography, "Gouverneur Kemble."

5. *The Foundry at Cold Spring*

1. Blake, *History of Putnam County*, p. 244; and Rutsch, *The West Point Foundry Site*, p. 39.

2. M. Wilson, *Thirty Years of Early His-*

tory of Cold Spring and Vicinity with Incidents, pp. 15–16.

3. Willis, *American Scenery,* p. 353.

4. *Putnam County History Workshop,* 1955.

5. Pelletreau, *History of Putnam County, New York,* p. 560.

6. "Gouverneur Kemble and West Point Foundry," *Americana,* 30:467.

7. Bruce, *The Hudson,* p. 129.

8. O'Brien, *American Sublime,* p. 158–161.

9. The quote is Lossing, *The Hudson from Wilderness to the Sea,* p. 246; the figure of 1400 men is from "A Cold Spring Walking Tour."

10. Van Zandt, *Chronicles of the Hudson,* p. 202 quoting Kemble *Journal of a Residence in America.*

11. Lossing *The Hudson from Wilderness to the Sea,* p. 246.

12. Kasson, *Civilizing the Machine,* pp. 168–171.

13. McShine, p. 63, quoting de Toqueville, *Journey to America* (1831).

6. *Mountain Spa*

1. Forman, "From Our Past," December 2, 1982.

2. Willis *Outdoors at Idlewild,* pp. v, 399.

3. Ibid., pp. 23–24.

4. Quotes from Holmes and Lowell are from *Dictionary of American Biography,* p. 306.

5. Willis, *Outdoors at Idlewild,* p. 236, 238, 244.

6. Beach, *Cornwall,* p. 42–44.

7. Willis, *Outdoors at Idlewild,* p. 21.

8. Beach, *Cornwall,* p. 42–44.

9. Ibid., p. 154.

10. O'Brien, *American Sublime,* p. 229.

11. Beach, *Cornwall,* p. 43.

12. Willis, *Outdoors at Idlewild,* p. 46.

13. Warner, *Hills of the Shatemuc,* p. 69.

14. O'Brien, *American Sublime,* p. 198.

15. Lossing, *The Hudson,* pp. 252–256.

16. Beach, *Cornwall,* p. 66.

17. Roe, *Play and Profit in My Garden,* p. 43.

18. Willis, *Outdoors at Idlewild,* pp. 141, 48.

19. Ibid., p. 47.

7. *The Landscape Garden*

1. "American Highland Scenery: Beacon Hill," *New York Mirror* (March 14, 1835) 12(37):293–294.

2. Curtis, "Memoir," in A.J. Downing *Rural Essays* pp. xxxii–xxxiii.

3. Hedrick, *A History of Horticulture in America to 1860,* p. 486.

4. Downing, *A Treatise on the Theory and Practice of Landscape Gardening,* pp. 23, 28–30, viii.

5. Ibid., pp. 60, 59, 328.

6. Ibid., p. 555.

7. Carmer, *The Hudson,* pp. 234–235.

8. Huth, *Nature and the American: Three Centuries of Changing Attitudes,* pp. 122–124.

9. Schuyler, "'Gems of Home Beauty on a Small Scale': Aspects of the Victorian Garden in America, 1840–1870" *Catalogue of the Old Lancaster Antiques Show,* November, 1983, p. 26.

10. Carolyn Flaherty, "The Domestic Architecture of Downing," *The Old House Journal* (October 1974) 2(10):9

11. Mitchell, *Out-of-Town Places,* pp. 3–7 also quoted in Jackson, "First Interpreter of American Beauty: A.J. Downing and the Planned Landscape," *Landscape* (Winter 1952) 1(3):16.

12. quoted in Carmer, *The Hudson,* p. 249.

13. Neil Larsen, draft paper, untitled, on historic architecture of the Hudson Highlands; New York State Office of Parks, Recreation and Historic Preservation; 1981.

8. *Millionaire's Row*

1. "Stuyvesant Fish," Dictionary of American Biography pp. 402–403.

2. "Edward H. Harriman" Dictionary of American Biography p. 300.

3. Letter from Susan Hale to her sister

Lucretia July 29, 1884. Sophia Smith Collection, Smith College, Northampton, Mass. Courtesy of Olana State Historic Site.

4. Bigelow, *Seventy Summers*, pp. 47–48.

5. Mearns, "Birds of the Hudson Highlands," 12: 124, 13:86; 12:127–128.

6. Cook, "Memoirs" p. Cr8–9.

7. Ibid., p. HF31.

8. Ibid., p. HF34.

9. Ibid., p. HF1–3.

10. *Art Times,* Jan/Feb, 1989, p. 1.

11. Winkler, *The First Billion: The Stillmans and the National City Bank*, p. 52.

12. Letters from Frederic Church to William Henry Osborn, November. 4 and 9, 1868 (Courtesy of Olana State Historic Site, Hudson N.Y.).

13. Letter from Susan Hale to her sister Lucretia July, 29, 1884, Sophia Smith Collection, Smith College, Northampton, Mass. (Courtesy of Olana State Historic Site).

14. Fish, *1600–1914*, p. 166.

15. Gray, "The Architectural Development of West Point," p. 14.

16. Hungerford, *Men and Iron: The History of New York Central*, pp. 135–136.

17. Jervis, "The Hudson River Railroad: A Sketch of its History, and Prospective Influence on the Railway Movement," p. 3.

18. Hudson River Rail Road Company, "Statement Showing the Prospects of Business and the Importance of the Proposed Rail Road," p. 14.

19. Jervis, "The Hudson River Railroad: A Sketch of its History, and Prospective Influence on the Railway Movement," p. 11.

20. Ibid., p. 6.

21. Fitzsimons, *The Reminiscences of John B. Jervis: Engineer of the Old Croton*, p. 178; Blake, *History of Putnam County, N.Y.*, p. 166.

9. *The Great Outdoors*

1. American Scenic and Historic Preservation Society (hereafter ASHPS) *Scenic and Historic America*, p. 11.

2. Ibid., p. 12.

3. ASHPS; *Scenic and Historic America*, p. 11.

4. Howell, *The Hudson Highlands*, 1:2.

5. Howell, *Highland Notes*, no page number.

6. ASHPS, *Scenic and Historic America*, p. 15.

7. Palisades Interstate Park Commission, *Sixty Years of Park Cooperation*, p. 17; ASHPS, *Scenic and Historic America*, p. 16.

8. Palisades Interstate Park Commission, *Sixty Years of Park Cooperation*, p. 17–19; ASHPS, *Scenic and Historic America*, p. 16; O'Brien, *American Sublime*, p. 244.

9. Palisades Interstate Park Commission, *Sixty Years of Park Cooperation*, pp. 19–21.

10. Albert, Letter, June 16, 1908.

11. Howell, *Highland Notes*, entry dated May 1, 1909.

12. ASHPS, "Thirteenth Annual Report."

13. Ibid., p. 74.

14. Handwritten notes from collection of the ASHPS, New York Public Library.

15. ASHPS, *Scenic and Historic America*, 1930, p. 27.

16. Original petition from the personal collection of J.W. Aldrich.

17. Williams, Letter, December 30, 1907.

18. Albert, Letter, December 16, 1907.

19. Hudson-Fulton Celebration Commission; Fourth Annual Report, 2:1106–1108.

20. ASHPS, *Scenic and Historic America*, 1930, p. 27.

21. Hudson-Fulton Celebration Commission; Fourth Annual Report, 2: 1110–1111.

22. Kennan, *E.H. Harriman*, 2:341–342.

23. ASHPS, *Scenic and Historic America*, 1930, p. 29.

24. Quoted in Kennan, *E.H. Harriman*,

1:191; Muir, *Edward Henry Harriman*, p. 35–36

25. Quoted in Kennan, *E.H. Harriman*, 2:34.

26. Muir, *Edward Henry Harriman*, pp. 3–6; Kennan, *E.H. Harriman* 1:208.

27. Kennan, *E.H. Harriman*, 2:33.

28. Ibid., 2:31–32, 2:339; 2:339–340; 2:343.

29. Palisades Interstate Park Commission, *Sixty Years of Park Cooperation*, p. 31.

30. Ibid., p. 30.

10. *Highways, Bridges and Tunnels*

1. All information about the Hudson-Fulton Celebration is from the Fourth Annual Report of the NYS Hudson-Fulton Celebration Commission, vols. 1 and 2.

2. New York Public Library, *An Exhibition*, pp. 101–103.

3. Lazarus White records his experiences as Division Engineer in a book titled *The Catskill Water Supply of New York City: History, Location, Sub-Surface Investigations and Construction*. All subsequent references to White's observations are from this source. The pages are in order of appearance of the quotes: 95,101–3, 102,427, 417, iii.

4. Dodge and Hoke, "Hudson River Crossings of the Catskill Aqueduct," *Proceedings of the Municipal Engineers for 1910* quoted in White, *The Catskill Water Supply*, p. 103.

5. Donaldson, "The Catskill Aqueduct," in Ambulando, ed. *In the Hudson Highlands*, p. 104.

6. Schiff, "In Celebration Cornwall 200 Years," pp. 40–41.

7. Ambulando, *In the Hudson Highlands*, p. 105.

8. Howell, *The Hudson Highlands*, p. 153.

9. Schiff, "In Celebration Cornwall 200 Years," pp. 36–37.

10. Partridge, Edward, undated, untitled newsclip from a scrapbook at Bear Mountain State Park.

11. Whitney, *Bridges*, p. 195.

12. Roebling, "Construction of Parallel Wire Cables for Suspension Bridges," p. 5. This pamphlet uses construction of the Bear Mountain Bridge as a case study, and contains many illustrations.

13. Steinman, *The Place of the Engineer*, p. 34.

14. Roebling, "Construction of Parallel Wire Cables for Suspension Bridges," p. 22–23; Steinman, *The Place of the Engineer*, p. 33.

15. Moses, Robert, Chairman, NYS Council of Parks, Letter to Governor Herbert Lehman, dated February 8, 1940 from the files at Bear Mountain State Park.

16. Mulligan, John; "Bear Mt. Bridge—at age 50" Associated Press; undated, unidentified newsclip from scrapbook at Bear Moutain State Park.

11. *Dynamite, Depression, and Design with Nature*

1. John Maude, excerpts from *Visit to the Falls of Niagara in 1800*, reprinted in Van Zandt, *Chronicles of the Hudson*, p. 141.

2. O'Brien, *American Sublime*, pp. 243, 325.

3. Hudson River Conservation Society; *First Annual Report* p. 22–23.

4. American Scenic and Historic Preservation Society (hereafter ASHPS), *Scenic and Historic America*, pp. 28–29. This bulletin is a special retrospective issue on Hudson River Conservation and provided much of the information for this chapter.

5. "Annual Message: The Hudson River" reprinted in Hudson River Conservation Society; *First Annual Report* p. 7.

6. Hudson River Conservation Society; *First Annual Report*; pp. 21–23. Corning's speech is reproduced in full.

7. Ibid., pp. 24–25.

8. Membership and financial information was compiled from the Annual Reports of the HRCS.

9. ASHPS; *Scenic and Historic America*, May 1940, p. 20.

10. ASHPS; *Scenic and Historic America*, May 1940, pp. 11, 20.

12. *Storm King*

1. Scenic Hudson Preservation Conference; "Some Important Dates in Storm King History" June 26,1968.

2. *The New York Times*; September 27, 1962; p. 1.

3. Tucker, "Environmentalism and the Leisure Class;" *Harper's*; p. 51.

4. Consolidated Edison Company of New York; "The Cornwall Hydroelectric Project: Power and Beauty for Tomorrow;" no date (sometime after 1965).

5. Talbot, *Power Along the Hudson*, p. 84.

6. *New York Times;* May 22, 1963.

7. Talbot, *Power Along the Hudson*, p. 85.

8. *New York Times*; May 22, 1963.

9. Talbot, *Power Along the Hudson*, p. 97.

10. Carl Carmer quoted in Doty, "Initial Brief to the Hearing Examiner;" June 15, 1963; pp. 23–26.

11. LeBoeuf; "Brief for Applicant, Consolidated Edison Company of New York;" June 13, 1964; p. 6, pp. 14–16; and in Talbot, *Power Along the Hudson*, pp. 99–106.

12. *Life*; July 31, 1964.

13. Talbot, *Power Along the Hudson*, p. 108–110.

14. The *Cornwall Local*; September 10, 1964.

15. New York State Legislature; "Preliminary Report of the Joint Legislative Committee on Natural Resources on the Hudson River Valley and the Consolidated Edison Company Storm King Mountain Project;" Robert Watson Pomeroy, Chr.; February 16, 1965; pp. 1–2.

16. Federal Power Commission; "Opinion No. 452, Project No. 2338; Opinion and Order Issuing License and Reopening and Remanding Proceeding for Additional Evidence on the Location of the Primary Lines and Design of Fish Protective Facilities," March 9, 1965; p. 40.

17. Talbot, *Power Along the Hudson*, p. 115.

18. Ibid., p. 118.

19. Ibid., p. 138.

20. House of Representatives Bill No. H.R. 3012, 89th Congress; January 18, 1965, introduced by Congressman Ottinger.

21. Senator Jacob Javits, Speech, March 4, 1965 published in "The Hudson: A River in Peril," Congressional Record, Proceedings and Debates of the 89th Congress, First Session; United States Government Printing Office No. 778–886–98643; 1965; p. 2.

22. Senator Robert F. Kennedy; speech of March 4, 1965 published in "The Hudson: A River in Peril," Congressional Record; p. 1.

23. Congressman Richard Ottinger; speech of May 3, 1965 published in "The Hudson: A River in Peril," Congressional Record, p. 1.

24. *The Times Herald Record*; March 15, 1965.

25. U.S. Court of Appeals, Second Circuit,; *Scenic Hudson Preservation Conference v. Federal Power Commission*, 354 F.2d. 608 (2d. Cir. 1965); in Sandler and Schoenbrod, eds. The Hudson River Power Plant Settlement; pp. 55–67.

26. Hudson River Valley Commission; *The Hudson: Summary Report*; February 1, 1966; p. 15.

27. Scully, Testimony before the FPC, Project No. 2338.

28. Atkinson, Testimony before the FPC, Project No. 2338.

29. Pough; Testimony before the FPC, Project No. 2338.

30. Harrison, Testimony before the FPC, Project No. 2338.

31. Scenic Hudson Preservation Conference "Bulletin #25," November 1967.

32. *The Harvard Crimson*; February 17, 1972.

33. Carmer, "Hollowed-Out Ground," *American Heritage* (June 1967) 18(4):96.

34. *Newsweek*; March 5, 1969.

35. *Fortune*, 1965.

36. U.S. Court of Appeals, Second Circuit,; *Scenic Hudson Preservation Conference v. Federal Power Commission*, 354 F.2d. 608 (2d. Cir. 1965); in Sandler and Schoenbrod, eds. The Hudson River Power Plant Settlement; p. 67.

37. *New York Times*, February 9, 1965.

38. Central Hudson Gas and Electric Corp.; letter to the Public Service Commisssion; quoted in the *Peekskill Evening Star* undated newsclip (c. 1970).

39. U.S. Federal Power Commission; Opinion No. 584, Consolidated Edison Co. of New York, Project No. 2338; Opinion and Order Issuing License Under Part I of the Federal Power Act.

40. U.S. Court of Appeals, Second Circuit,; *Scenic Hudson Preservation Conference v. Federal Power Commission*, 453 F.2d. 463 (2d. Cir. 1971); in Sandler and Schoenbrod, eds. *The Hudson River Power Plant Settlement*; p. 92.

41. U.S. Court of Appeals, Second Circuit,; *Hudson River Fishermen's* Association v. Federal Power Commission, 498 F.2d. 827 (2d. Cir. 1974); in Sandler and Schoenbrod, eds. *The Hudson River Power Plant Settlement*; p. 97.

42. Luce, "Power for Tomorrow: The Siting Dilemma."

Bibliography

Adams, Arthur G. *The Hudson: A Guidebook to the River*. Albany: State University of New York Press, 1981.

Adams, Arthur G. *The Hudson River in Literature: An Anthology*. Albany, N.Y.: State University of New York Press, 1980.

Albert, F.P. Letter dated Dec. 16, 1907 to Francis Lynde Stetson, Chairman, Law Committee, Hudson-Fulton Celebration, from collection of the Hudson River Conservation Society

Albert, F.P. Letter dated June 16, 1908 to Eben E. Olcott from the collection of the Hudson River Conservation Society

Ambrose, Stephen E. *Duty, Honor, Country: A History of West Point*. Baltimore: John Hopkins Press, 1966.

Ambulando, Solvitur, ed. *In the Hudson Highlands*. New York: Appalachian Mountain Club, 1945.

"American Highland Scenery: Beacon Hill." *New York Mirror*, (March 14, 1835) 12:37.

American Scenic and Historic Preservation Society. "Mount Taurus Quarry to be Stopped." *Scenic and Historic America* (June 1931) 3:2.

American Scenic and Historic Preservation Society. *Scenic and Historic America* (May 1940) 5(3). Special bulletin on Hudson River Conservation. A joint publication of the American Scenic and Historic Preservation Society and the Hudson River Conservation Society.

American Scenic and Historic Preservation Society. "Thirteenth Annual Report." 1908.

American Scenic and Historic Preservation Society. *Scenic and Historic America* (June 1930) 2(2).

Americana, vol. 30. New York: The American Historical Society, 1936.

"Andrew Jackson Downing." *Dictionary of*

American Biography. Allen Johnson and Dumas Malone, eds. New York: Scribner, 1928–1964.

Beach, Lewis. *Cornwall*. New York: E. M. Ruttenber and Son, 1873.

Berg, Jean. "Commerce in Cold Spring 1776–1976." *The Putnam County News and Recorder*, June 30, 1976.

Berkey, Charles P. "Geology of the New York City (Catskill) Aqueduct." *Education Department Bulletin No. 489; New York State Museum Bulletin 146*. Albany: University of the State of New York, 1911.

Bigelow, Poultney. *Seventy Summers*. London: Edward Arnold and Co., 1925.

Blake, William J. *The History of Putnam County, N.Y.* New York: Baker and Scribner, 1849.

Board of Water Supply of the City of New York. *Catskill Water Supply: A General Description*. New York, 1928.

Boyle, Robert H. *The Hudson River: A Natural and Unnatural History*. New York: Norton, expanded edition, 1979.

Bradley, Lt. Col. John H. *West Point and the Hudson Highlands During the American Revolution* [Brochure]. West Point: United States Military Academy, 1976.

Bruce, Wallace. *The Hudson*. New York: Bryant Literary Union, 1894.

Bruce, Wallace. *The Hudson: Three Centuries of History, Romance, and Invention*. New York: Bryant Union Co., 1907 (centennial edition).

Bruce, Wallace. *The Hudson by Daylight by Thursty McQuill*. New York: Gaylord Watson, 1873.

Bryant, William Cullen. *Thanatopsis*. Boston: The Bibliophile Society, 1927 (originally published 1817).

Buckingham, J. S. *America, Historical, Statistic, and Descriptive*, vol. 2. London: Fisher, Son, and Co., 1841.

Calihan,Robert J. and Judith LaBelle Larsen. "The Hudson Highlands." Poughkeepsie: Scenic Hudson Preservation Conference, 1981 (unpublished manuscript).

Callow, James T. *Kindred Spirits: Knickerbocker Writers and American Artists 1807–1855*. Chapel Hill: University of North Carolina Press, 1967.

Canfield, Cass. *The Incredible Pierpont Morgan: Financier and Art Collector*. New York: Harper and Row, 1974.

Carmer, Carl. "Hollowed-Out Ground." *American Heritage*, (June 1967) 18(4).

Carmer, Carl. *The Hudson*. New York: Rinehart, 1939, 1974 edition.

Carroll, Charles. "Journal of Charles Carroll of Carrollton, During His Visit to Canada in 1776. As One of the Commissioners from Congress; With a Memoir and Notes." In Roland Van Zandt, ed. *Chronicles of the Hudson: Three Centuries of Travellers' Accounts*. New Brunswick, NJ: Rutgers University Press, 1971.

Clapp, Margaret. *Forgotten First Citizen: John Bigelow*. Boston: Little, Brown and Co., 1947.

Clinton, Sir Henry. *The American Rebellion*. William B. Willcox, ed. New Haven, CT: Yale University Press, 1954.

A Cold Spring Walking Tour. Uncopyrighted pamphlet produced by the Cold Spring Area Chamber of Commerce, The Putnam County Historical Society, and the Village of Cold Spring Historic Preservation Board.

Colyer, William H. *Sketches of the North River*. New York: author, 1838.

Cook, Grace. "Memoirs" (unpublished mimeograph courtesy of Anne and Consantine Sidamon-Eristoff, Highland Falls, N.Y.)

Cooper, James Fenimore. *The Spy: A Tale of the Neutral Ground*. London: George Routledge and Sons, Ltd., 1849.

Cornwall Local, September 10, 1964. Cornwall, N.Y.

Crane, John, and James F. Kieley. *West Point: The Key to America*. New York: McGraw Hill Book Company, 1947.

Cummings, Thomas S. *Historic Annals of the National Academy of Design*. Philadelphia: George W. Childs, 1865.

Curtis, George William. "Memoir." In A.J. Downing, *Rural Essays*. New York: Geo. A. Leavitt, 1869.

D'Auermont, Frances Wright. *Society and*

Manners in America in a Series of Letters from that Country to a Friend in England by an Englishwoman. N.Y.: E. Bliss and White, 1821.

DeLanoy, Nelson. "From Our Past," A series of articles appearing in the *Putnam County News and Recorder*

DeLanoy, Nelson, and Marguerite W. Rogers. *George Pope Morris: 1802–1864*. Cold Spring, N.Y.: Putnam County Historical Society, 1964.

Dickens, Charles. *American Notes for General Circulation*. London: Penguin Books, 1972 (reprint of the 1842 edition).

Dictionary of American Biography, Allen Johnson and Dumas Malone, eds. New York: Scribner, 1928–1964.

Doty, Dale. *Initial Brief to the Hearing Examiner of the Scenic Hudson Preservation Conference*, United States of America before the Federal Power Commission in the matter of Consolidated Edison Company of New York, Inc. Project No. 2338, June 15, 1964.

Downing, A.J. *Cottage Residences*. New York: Wiley and Putnam, 1844.

Downing, A.J. *Rural Essays*. George William Curtis, ed. New York: Leavitt and Allen, 1857.

Downing, A. J. *A Treatise on the Theory and Practice of Landscape Gardening Adapted to North America With a View to the Improvement of Country Residences*. Seventh edition with a Supplement by Henry Winthrop Sargent. New York: Orange Judd Agricultural Book Publisher, 1865.

Drake, Joseph Rodman. *The Culprit Fay*. New York: Rudd and Carlton, 1859.

Dwight, Timothy. *Travels in New England and New York*, vol.3 Barbara Miller Solomon, ed. Cambridge: Belknap Press of Harvard University Press, 1969 (originally published New Haven: S. Converse, 1822).

Erickson, Margery O. *A Few Citizens of Philipstown*. Garrison, N.Y.: Capriole Press, 1990.

Evers, Alf. *The Catskills from Wilderness to Woodstock*. Garden City, New York: Doubleday, 1972.

Federal Power Commission. Opinion No. 452, Project No. 2338; Opinion and Order Issuing License and Reopening and Remanding Proceeding for Additional Evidence on the Location of the Primary Lines and Design of Fish Protective Facilities, March 9, 1965.

Fitzsimons, Neal, ed. *The Reminiscences of John B. Jervis: Engineer of the Old Croton*. Syracuse, N.Y.: Syracuse University Press, 1971

Fitzpatrick, John C., ed. *The Writings of George Washington From Original Manuscript Sources 1745–1799*. vols.4–5. Washington D.C.: U.S. Government Printing Office, 1932.

Flaherty, Carolyn. "The Domestic Architecture of Downing," *The Old House Journal*, October 1974. vol.2, no.10.

Force, Peter,ed. *American Archives*, Fourth and Fifth Series. Washington, D.C.: 1837–1853.

Ford, Worthington Chauncey, ed. *Journals of the Continental Congress 1774–1789*. Washington D.C.: Government Printing Office, 1905.

Forman, Sidney. *The American Revolution in the Hudson Highlands*. Highlands, N.Y.: author, 1982.

Forman, Sidney. "From Our Past." *News of the Highlands*, Town of Highlands, N.Y., December 1982.

Forman, Sidney. *West Point: A History of the United States Military Academy*. New York: Columbia University Press, 1950. *Fortune*, 1965.

Gerwig, Henrietta. *Historic Philipstown in the Highlands of the Hudson*. (pamphlet)

Giles, Dorothy. *A Brief Story of the American Revolution in the Hudson Highlands*. Cold Spring, N.Y.: Putnam County Historical Society, 1976.

Giles, Dorothy. "Marshmoor and Undercliff." *Putnam County History Workshop*, Mimeograph, 1955 (Putnam County Historical Society files).

Graham, Frank. *The Adirondack Park: A Political History*. New York: Alfred A. Knopf, 1978.

Gray, Col. David W. "The Architectural Development of West Point." Department of Military Topography and graphics, U.S. Military Academy, West Point, N.Y., 1951 (unpublished paper).

The Harvard Crimson, February 17, 1972. Cambridge: Harvard University

Hastings, Hugh, ed. *Public Papers of George Clinton*, vol.1. Albany: Published by the State of New York as the third annual report of the State Historian, 1899.

Hedrick, U.P. *A History of Horticulture in America to 1860*. New York: Oxford University Press, 1950.

Hillery, Horace. "Putnam County in the Civil War." Putnam County Historian, Patterson, N.Y.: 1961.

Howat, John K. *The Hudson River and Its Painters*. New York: Penguin Books, 1978 (copyright Scenic Hudson Preservation Conference, Viking, 1972).

Howell, William Thompson. *The Hudson Highlands*. 2 vols. under one cover. New York: Walking News, Inc., 1982. (reprinted from 1933–34 uncopyrighted editions)

Howell, William Thompson. *Highland Notes*. Uncopyrighted diary/scrapbook 2 vols. c. 1910 (New York Public Library).

Hudson River Conservation Society. *Certificate of Incorporation*, 1936.

Hudson River Conservation Society. *First Annual Report 1937*. West Point, N.Y., 1937.

Hudson River Conservation Society. *Second Annual Report 1937*. West Point, N.Y., 1938.

Hudson River Fishermen's Association vs. Federal Power Commission, 498 F. 2d 463 (2d Cir 1971)

Hudson River Rail Road Company. "Statement Showing the Prospects of Business and the Importance of the Proposed Rail Road." New York: E.B. Clayton and Sons, Stationers, 1846.

Hungerford, Edward. *Men and Iron: The History of New York Central*. New York: Thomas Y. Crowell, 1938.

Hunt, Freeman. *Letters about the Hudson and Its Vicinity, Written in 1835 and 1836*. New York: Freeman Hunt and Co., 1836.

Huth, Hans. *Nature and the American: Three Centuries of Changing Attitudes*. Los Angeles: University of California Press, 1957.

Irving, Washington. *Bracebridge Hall or the Humorists*. Surrey edition, vol.2. New York: G.P. Putnam's Sons, 1896.

Irving, Washington. *The Crayon Miscellany, The Complete Works of Washington Irving*, vol. 22, Dahlia Kirby Terrell, ed. Boston, MA: Twayne Publishers, 1979.

Irving, Washington. *Knickerbocker's New York: A History of New York from the Beginning of the World to the End of the Dutch Dynasty by Diedrich Knickerbocker*. New York: G. P. Putnam's Sons, 1888 (originally published 1809).

Jackson, W.G. "First Interpreter of American Beauty: A.J. Downing and the Planned Landscape." *Landscape*, Winter 1952, vol.1, no.3.

Jensen, Oliver. *The American Heritage History of Railroads in America*. New York: American Heritage, 1975.

Jervis, John B. "The Hudson River Railroad: A Sketch of its History, and Prospective Influence on the Railway Movement." (reprinted from *Hunt's Merchant's Magazine*, March 1850).

Jervis, John B. "Report on the Project of a Railroad on the East Bank of the Hudson River, from New York to Albany." New York, January 20, 1846 (no copyright).

Kasson, John F. *Civilizing the Machine: Technology and Republican Values in America, 1776–1900*. New York: Grossman Publishers, 1976.

Kennan, George. *E.H. Harriman: A Biography*. 2 vols. Freeport, N.Y.: Books for Libraries Press, 1922, reprinted in 1969.

Lancaster, Bruce. *The American Heritage Book of the Revolution*. New York: American Heritage, 1971

Larsen, Neil. Draft paper, untitled, on historic architecture of the Hudson Highlands. New York State Office of Parks, Recreation and Historic Preservation, 1981.

LeBoeuf, Randall. Brief for Applicant, Consolidated Edison Company of New York, June 13, 1964.

Leslie, Eliza. "Recollections of West Point."

Graham's Magazine, April 1842, pp. 207–209,290–295.

Life, July 31, 1964.

Limburg, K. E., M.A. Moran, W.T. McDowell, *The Hudson River Ecosystem*, New York: Springer Verlag, 1986

Lossing, Benson. *The Hudson from the Wilderness to the Sea.* New York: Virtue and Yorston, 1866.

Luce, Charles F. "Power for Tomorrow: The Siting Dilemma." Remarks before the Association of the Bar of the City of New York, November 18, 1969.

Maas, John. *The Gingerbread Age: A View of Victorian America.* New York: Rinehart, 1957.

Mabie, Hamilton Wright. *The Writers of Knickerbocker New York.* New York: The Grolier Club, 1912.

Marryat, Frederick. *A Diary in America With Remarks on its Institutions.* New York: Alfred A. Knopf, 1962.

McCrea, Tully. *Dear Belle: Letters from a Cadet to his Sweetheart,* 1858–1865. Middletown, CT: Wesleyan University Press, 1965.

McEntee, Girard L. "Andrew Jackson Downing, Landscape Architect." Publication No. 36, The Historical Society of Newburgh Bay and the Highlands, Newburgh, N.Y., 1950.

McPhee, John. *In Suspect Terrain.* New York: Farrar, Straus, Giroux, 1982, 1983.

McShine, Kynaston, ed. *The Natural Paradise: Painting in America 1800–1950.* New York: Museum of Modern Art, 1976.

Mearns, Dr. Edgar A. "Birds of the Hudson Highlands." *Bulletin of the Essex Institute*, vols. 10–13, 1878–1881. Salem, Mass.: Salem Press.

Mearns, Dr. Edgar. "A Study of the Vertebrate Fauna of the Hudson Highlands, with Observations on the Mollusca, Crustacea, Lepidoptera and the Flora of the Region." *Bulletin of the American Museum of Natural History*, vol. 10, 1898.

Melville, Herman. *Moby Dick or, The Whale.* Berkeley: University of California Press, 1979.

Mercer, Lloyd J. *E. H. Harriman: Master Railroader.* Boston: Twayne Publishers, 1985.

Milbert, J. *Picturesque Itinerary of the Hudson River and Peripheral Parts of North America.* Upper Saddle River, NJ: The Gregg Press, 1969 (translated from the 1828–1829 French edition by Constance Sherman).

Mitchell, Donald G. *Out-of-Town Places.* New York: Charles Scribner's Sons, 1907.

Montule, Edouard de. *Travels in America, 1816–1817.* Bloomington: Indiana University Press, Social Science Series #9, 1950. (Translated from the 1921 Paris edition by Edward Seeber.)

Moore, Frank and John Anthony Scott, ed. *The Diary of the American Revolution 1775–1781.* New York: Washington Square Press, 1967.

Muir, John. *Edward Henry Harriman.* Garden City, N.Y.: Doubleday, Page and Co., 1912.

Mylod, John. *Biography of a River, the People and Legends of the Hudson Valley.* New York: Hawthorne, 1969.

Nash, Roderick. *The American Environment.* Reading, Mass.: Addison-Wesley Publishing Co., 1976.

Nash, Roderick. *Wilderness and the American Mind.* New Haven, CT: Yale University Press, 1967, third edition 1982.

Newsweek, March 5, 1969.

Newton, Norman T. *Design on the Land: The Development of Landscape Architecture.* Cambridge: The Belknap Press of Harvard University Press, 1971.

New York Public Library. *An Exhibition Illustrating the History of the Water Supply of the City of New York from 1639 to 1917*, held at the New York Public Library May 1 to November 6, 1917. New York Public Library, 1917 (pamphlet).

New York State. Hudson River Valley Commission. *The Hudson: Summary Report* Albany, N.Y., February 1, 1966

New York State. *Preliminary Report of the Hudson Valley Survey Commission to the Legislature,* Legislative Document no. 98. Albany, N.Y., March 18, 1938.

New York State. *Report of the Hudson Valley*

Survey Commission to the Legislature, Legislative Document no. 71. Albany, N.Y., March 1, 1939.

New York State Education Department. *The Hudson Valley and the American Revolution*. Albany, N.Y.: New York State Historic Trust, 1968.

New York State Hudson-Fulton Celebration Commission. *The Hudson Fulton Celebration 1909*. Fourth Annual Report to the Legislature of the State of New York, 2 vols., prepared by Edw. Hagaman Hall, Albany, N.Y. 1910.

New York Legislature. "An Act to incorporate the Bear Mountain Hudson River Bridge Company, and to authorize the construction of a bridge across the Hudson river near the village of Peekskill, together with approaches thereto and to define the rights of the state respecting such bridge." Senate Bill No.626. Albany, N.Y. February 8, 1922.

New York State Legislature. "Preliminary Report of the Joint Legislative Committee on Natural Resources of the Hudson River and the Consolidated Edison Company Storm King Mountain Project." Robert Watson Pomeroy, Chr., February 16, 1965.

New York State Museum and Science Service. *Geology of New York: A Short Account*. Albany, N.Y.: University of the State of New York/State Education Department, 1966.

New York Times, February 9, 1965.

Noble, Louis L. *The Life and Works of Thomas Cole*. Elliot S. Vesell, ed. Cambridge: Belknap Press of Harvard University Press, 1964.

Novotny, Ann. "Research Reports." prepared for Selvage and Lee, 500 Fifth Ave., New York on behalf of Scenic Hudson Preservation Conference, July 30, 1964.

O'Brien, Raymond J. *American Sublime: Landscape and Scenery of the Lower Hudson Valley*. New York: Columbia University Press, 1981.

Palisades Interstate Park Commission. *Sixty Years of Park Cooperation: New York-New Jersey-A History 1900–1960*. Bear Mountain State Park, Bear Mountain, N.Y., 1960 (pamphlet).

Palmer, Lt. Col. Dave Richard. *The River and the Rock: The History of Fortress West Point, 1775–1783*. New York: Greenwood Publishing Co., 1969.

Partridge, Dr. Edward. "The New Storm King Highway." December 15, 1921, unidentified newsclip. (Franklin D. Roosevelt Library)

Pelletreau, William S. *History of Putnam County, New York*. Philadelphia, PA: W. W. Preston and Co., 1886.

Poe, Edgar Allan; "The Domain of Arnheim or the Landscape Garden," *Best Known Works of Edgar Allan Poe*. Garden City, N.Y.: Blue Ribbon Books, 1941.

Putnam County Historical Society. "Putnam County History Workshops," mimeographs 1955, 1957.

Ransom, James. *Vanishing Ironworks of the Ramapoes*. New Brunswick, NJ: Rutgers University Press, 1966.

Raup, Hugh M. "Botanical Studies in the Black Rock Forest." *Black Rock Forest Papers*, Bulletin no.7, 1938, Cornwall, N.Y.

Reeder, Col. Red. *Heroes and Leaders of West Point*. New York: Thomas Nelson, Inc., 1970.

Roberts, Robert B. *New York's Forts in the Revolution*. Rutherford, NJ: Fairleigh Dickinson University Press, 1980.

Roe, Rev. E. P. *Play and Profit in My Garden*. New York: Dodd and Mead, 1873.

Roebling Co. "Construction of Parallel Wire Cables for Suspension Bridges." John A. Roebling's Sons Company, Trenton, NJ, 1925 (pamphlet).

Rutsch, Edward; JoAnn Cotz, Brian Morrell, Herbert Githens and Leonard Eisenberg. *The West Point Foundry Site: Cold Spring* (Putnam County, N.Y.). Newton, NJ: Cultural Resource Management Service. Ralph Brill Project Director; copyright 1979 by Ralph Brill Associates.

Sanders, John. *Principles of Physical Geol-*

ogy. New York: John Wiley and Sons, 1981.

Sandler, Ross and David Schoenbrod, eds. *The Hudson River Power Plant Settlement: Conference Materials.* New York University School of Law, December 1981.

Satterlee, Herbert L.; *J. Pierpont Morgan: An Intimate Portrait*. New York: Macmillan, 1939.

Saunders, Jean. Garrison's Landing, pamphlet, privately published, 1966

Scenic Hudson Preservation Conference, "Some Important Dates in Storm King History." (Flyer).

Scenic Hudson Preservation Conference, *Bulletin #25*, November 1967.

Schiff, Martha, ed. "In Celebration Cornwall 200 Years." *The Cornwall LOCAL*, Cornwall, N.Y.: News of the Highlands, Inc., 1976.

Schneider, David and Harry Elmer Barnes. "The Rise of Human Institutions." *History of the State of New York,* vol.8. New York: Columbia University Press, 1935.

Schuberth, Christopher J. *Geology of New York City and Environs*. Garden City, N.Y.: Natural History Press, 1968.

Schuyler, David. "'Gems of Home Beauty on a Small Scale': Aspects of the Victorian Garden in America, 1840–1870." *Catalogue of the Old Lancaster Antiques Show,* November, 1983.

Simpson, Jeffrey. *Officers and Gentlemen: Historic West Point in Photographs*. Tarrytown, N.Y.: Sleepy Hollow Press, 1982 (copyright Sleepy Hollow Restorations).

Smith, Sydney. *Edinburgh Review*. (Jan. 1820), vol. 33,

Stedman, Charles. *History of the Origin, Progress and Termination of the American War*. 2 vols. Printed for Mssrs. P. Wogan, P. Byrne, J. Moore, and W. Jones. Dublin, Ireland: 1794.

Steinman, Dr. D.B. "The Place of the Engineer in Civilization." State College Record, University of North Carolina, School of Engineering, Raleigh, NC, 1939.

Sutcliffe, Alice Crary. *Robert Fulton and the "Clermont*'.' New York: The Century Co., 1909.

Talbot, Allan. *Power Along the Hudson; The Storm King Case and the Birth of Environmentalism*. New York: E. P. Dutton and Co., 1972.

Thompson, H.D. "Hudson Gorge in the Highlands." *Bulletin of the Geological Society of America*, December 1936, no.47, pp.1831–1848.

Transactions of the American Climatological Association, vol. 4, 1887.

Tucker, William E. "Environmentalism and the Leisure Class." *Harper's*, December 1977.

U.S. Congress. "The Hudson: A River in Peril." *Congressional Record*, Proceedings and Debates of the 89th Congress, First Session. United States Government Printing Office No.778–886–98643, 1965.

U.S. Court of Appeals, Second Circuit. *Scenic Hudson Preservation Conference v. Federal Power Commission*, 354 F.2d. 608 (2d. Cir. 1965), in Sandler and Schoenbrod, eds. *The Hudson River Power Plant Settlement: Conference Materials,* N.Y.U School of Law, December 1981.

U.S. Department of the Interior, National Park Service. National Register of Historic Places Inventory—Nomination Forms, "Castle Rock," Dec. 12, 1977.

U.S. Department of the Interior, National Park Service. National Register of Historic Places Inventory—Nomination Forms, "Hudson Highlands Multiple Resource Listing" 1977–1985 (N.Y.S Office of Parks, Recreation and Historic Preservation, Albany, N.Y.)

U.S. Federal Power Commission. Testimony before the Federal Power Commission in the Matter of the Application for Construction of Hydroelectric Plant at Cornwall, New York by Consolidated Edison Co. of New York, Project No. 2338

U.S. House of Representatives. Bill No. H.R. 3012, 89th Congress, January 18, 1965, introduced by Congressman Richard Ottinger.

Van Zandt, Roland, ed. *Chronicles of the Hudson: Three Centuries of Travellers' Accounts*. New Brunswick, NJ: Rutgers University Press, 1971.

Vaux, Calvert. *Villas and Cottages*. New York: Harper and Brothers, Publishers, 1857.

Venables, Robert. *The Hudson Valley in the American Revolution*. Albany: N.Y.S American Revolution Bicentennial Commission, 1975.

Visitor's Guide to a Community that Lives in History: Cold Spring, Nelsonville, Garrison-on-the-Hudson, New York. Cold Spring, N.Y.: Cold Spring Area Chamber of Commerce, 1980. (pamphlet)

Warner, Susan. *The Hills of the Shatemuc*. Philadelphia: J.B. Lippincott Co., 1856.

White, Lazarus. *The Catskill Water Supply of New York City: History, Location, Sub-Surface Investigations and Construction*. New York: Wiley, 1913.

Whitney, Charles S. *Bridges: A Study in Their Art, Science and Evolution*. New York: William Edward Rudge, Publisher, 1929.

Wilkie, Richard. *The Illustrated Hudson River Pilot*. Albany: Three City Press, 1974.

Willis, Nathaniel P. *American Scenery*. Barre, Mass: Imprint Society, 1971. (Reprint 2 vols. London: Geo. Virtue, 1840).

Willis, Nathaniel P. *Outdoors at Idlewild or The Shaping of a Home on the Banks of the Hudson*. New York: Charles Scribner, 1855.

Wilson, M. *Thirty Years of History of Cold Spring and Vicinity, With Incidents By One Who Has Been A Resident Since 1819*. Cold Spring, N.Y.: Shram Printing House, 1886

Winkler, John K. *The First Billion: The Stillmans and the National City Bank*. New York: The Vanguard Press, 1934.

Winkler, John K. *Morgan the Magnificent: The Life of J. Pierpont Morgan (1837–1913)*. Garden City, N.Y.: Garden City Publishing Co.

Wyckoff, Jerome. *Rock Scenery of the Hudson Highlands and Palisades*. Glens Falls, N.Y.: Adirondack Mountain Club, 1971.

Index

Note: Numbers that appear in *italics* refer to captions of illustrations.